"Content strategy is a growing practice across many fields. This book is an excellent introduction to content strategy, particularly for professional and technical communicators. It provides a bird's-eye view of everything content, so that the learner understands the value of content strategy in a broader context. It also gives a worm's-eye view of the skills and steps involved in creating compelling content. The many practical examples and useful exercises help set this book apart."

—**Dr. Quan Zhou**, *Professor, Department Chair, Metropolitan State University*

"This much-needed text is the first in the field of technical communication to articulate—with teachers and students in mind—the practices, genres, workflows, and underlying principles of the emerging discipline of content strategy. Teachers and students will find valuable guidance on how to approach different stages of the content strategy process, from conducting a content audit to developing content models to delivering, governing, and maintaining genres. Written in an accessible and engaging style and full of robust examples and downloadable templates, this text is a must-read for anyone interested in learning about content strategy and developing skills for content strategy work, which spans across disciplines and organizational contexts, large and small."

—**Dr. Rebekka Andersen**, *Associate Professor, Associate Director of Writing in the Professions, University of California, Davis*

"Guiseppe Getto, Jack Labriola, and Sheryl Ruszkiewicz's new textbook on content strategy fills a significant hole in the field of technical and professional technical communication. If you're like me and teach whole courses dedicated to content strategy, you already know that there are many books for marketing or public relations students, but those devoted to TPC students are difficult to find. They're even more difficult to find when you're looking for authors who are experts in the field with real dirt under their fingernails gained through actual practice. I'm really looking forward to using this book in my next content strategy course instead of having to cobble together selections from disparate publications."

—**Dr. Tharon Howard**, *Professor, MA in Professional Communication Graduate Program Director, Clemson University*

Content Strategy

This comprehensive text provides a how-to guide for content strategy, enabling students and professionals to understand and master the skills needed to develop and manage technical content in a range of professional contexts.

The landscape of technical communication has been revolutionized by emerging technologies such as content management systems, open-source information architecture, and application programming interfaces that change the ways professionals create, edit, manage, and deliver content. This textbook helps students and professionals develop relevant skills for this changing marketplace. It takes readers through essential skills including audience analysis; content auditing; assembling content strategy plans; collaborating with other content developers; identifying appropriate channels of communication; and designing, delivering, and maintaining genres appropriate to those channels. It contains knowledge and best practices gleaned from decades of research and practice in content strategy and provides its audience with a thorough introductory text in this essential area.

Content Strategy works as a core or supplemental textbook for undergraduate and graduate classes, as well as certification courses, in content strategy, content management, and technical communication. It also provides an accessible introduction for professionals looking to develop their skills and knowledge.

Guiseppe Getto is Associate Professor of Technical Communication at Mercer University and President and Founder of Content Garden, Inc., a content strategy and UX consulting firm.

Jack T. Labriola is Experience Design Senior Researcher at Truist and is Vice President of User Experience and Content Strategy at Content Garden, Inc.

Sheryl Ruszkiewicz is Special Lecturer at Oakland University, where she teaches in the First-Year Writing Program.

ATTW Series in Technical and Professional Communication
Tharon Howard, Series Editor

For additional information on this series please visit www.routledge.com/ATTW-Series-in-Technical-and-Professional-Communication/book-series/ATTW, and for information on other Routledge titles visit www.routledge.com.

Content Strategy

A How-to Guide

Guiseppe Getto
Jack T. Labriola
Sheryl Ruszkiewicz

Routledge
Taylor & Francis Group

NEW YORK AND LONDON

Cover image: © Getty Images

First published 2023
by Routledge
605 Third Avenue, New York, NY 10158

and by Routledge
4 Park Square, Milton Park, Abingdon, Oxon, OX14 4RN

Routledge is an imprint of the Taylor & Francis Group, an informa business

Library of Congress Cataloging-in-Publication Data
Names: Getto, Guiseppe (Associate professor of technical
    communication), author. | Labriola, Jack T., author. |
    Ruszkiewicz, Sheryl, author.
Title: Content strategy: a how-to guide/Guiseppe Getto,
    Jack T. Labriola, Sheryl Ruszkiewicz.
Identifiers: LCCN 2022019268 (print) | LCCN 2022019269 (ebook) |
    ISBN 9780367751036 (paperback) | ISBN 9780367759506 (hardback) |
    ISBN 9781003164807 (ebook)
Subjects: LCSH: Business planning. | Business communication—Planning. |
    Branding (Marketing)
Classification: LCC HD30.28. G4818 2023 (print) | LCC HD30.28 (ebook) |
    DDC 658.4/012–dc23/eng/20220510
LC record available at https://lccn.loc.gov/2022019268
LC ebook record available at https://lccn.loc.gov/2022019269

ISBN: 9780367759506 (hbk)
ISBN: 9780367751036 (pbk)
ISBN: 9781003164807 (ebk)

DOI: [10.4324/9781003164807]

Typeset in Bembo
by Apex CoVantage, LLC

For all the content people out there, new and experienced. You know who you are. May this book make your lives a little easier.

Contents

Introduction

What Is Content Strategy?

Content strategy is a wide-ranging field that encompasses everything from people who manage websites for a living to people who create repositories of content that companies can use for documentation. As organizations across industries, including technology, engineering, healthcare, and education, have grown, so have their needs for various types of content.

By "content" we mean *useful information that an audience will consume*. If you are a cook who keeps track of recipes in a private notebook, you probably don't think of yourself as a content strategist. If you help a food manufacturing company to manage its public information across a variety of demographics and distribution channels, however, you might, in fact, be a content strategist.

People who manage content for a living are everywhere these days. Besides content strategists, they include:

- Technical writers
- Technical editors
- Journalists
- Freelance writers
- Bloggers
- Creative writers
- Educators
- Instructional designers
- Researchers
- Publishers
- Website managers

Jobs like these have grown along with a global economy that involves processing more and more information. Many of us wake up to a smart device that serves as our alarm clock, personal digital assistant (PDA), and messaging system. We check updates on our laptop via social media platforms, company websites, online magazines, and news websites. We watch short videos, TV shows, and films on our smart television via streaming websites like YouTube, Netflix, and Hulu. We shop for products ranging from food to furniture through e-commerce websites owned by both large and small companies. We even use the websites of national and regional non-profits to connect us to communities and individuals in need. And many of us also work to develop, manage, or distribute information to one or more audiences, whether we are a teacher, a communications manager for a non-profit, a documentation specialist within a software company, or even a lawyer.

All of these forms of information have to be managed within their specific channels, defined as *any means where content is distributed in order for it to be consumed by a specific audience.* The same piece of information can't seamlessly flow through a blog displayed on a large desktop monitor, a notification on a small smart device, and a streaming video watched on a digital advertising display. Someone has to develop, format, edit, style, and deliver the content to each of those channels. If they are very savvy and have access to the best tools available, they may be able to save time by creating the content in a format that will be styled by each channel appropriately. Sometimes, this is impossible, such as when delivering the same information via a written genre like a technical report and a videographic genre like a television newscast.

Because of these challenges, organizations that have funding to do so, and that know such professionals exist, hire dedicated professionals who can help them manage all of their information for all of their channels. If they are a larger organization, they probably hire multiple someones. Within even small-to-medium-sized organizations, professionals such as these are responsible for producing and managing such varying genres of content as:

- Internal reports
- Emails
- Memos
- Handbooks
- Strategy plans
- Webpages
- Ebooks
- Content repositories such as content management systems
- Social media posts

These professionals aren't always called content strategists, but essentially that's what they are. What sets them apart from the professionals we mentioned earlier, such as technical writers or educators, you may ask?

The answer to that question is really a core purpose of this book. We want to make it clear to people new to content strategy what this field is, what people in it do for a living, and how people can use its best practices in any career that involves communication.

Keep in mind: within all these organizations we've alluded to so far, within universities, and K-12 schools, and news companies, and law offices, there are professionals working to develop and publish content. These folks typically have some kind of communication-related verb in their title such as "writer," "editor," or "manager." They're typically responsible for a *specific type of content*. A technical writer working for a software company is most likely in charge of writing software documentation. An editor for a small publishing company is typically responsible for editing manuscripts that the company receives from its authors. A website manager is typically responsible for managing content within a single website, or sometimes even just a very complex *part* of a single website.

But who is managing all of this content collectively within these organizations? Who is making sure it all makes sense and aligns with an organization's goals? Who is making sure that it's appropriate to every channel an audience uses to consume it? Who is making sure it's updated, optimized, and delivered on time?

Enter the role of the content strategist as purveyor and manager of relevant information. This is a person who is not a manager, a writer, or an editor, per se, but who often works in close proximity with these other types of professionals to make sure content is useful and usable for its target audiences. Content strategists will often conduct focus groups or other research to adapt existing content to customer needs. Overall, they manage, shape, and deliver content and teach organizations, and people within organizations, how to do so as well.

And there are several books on content strategy that have been written over the past several decades since the field emerged. Some of them are listed in Chapter 1 of this book as examples of further reading. Few of them serve as introductory texts for content strategy, however, which is where this book comes in.

This book is for:

- Undergraduate students who are interested in a brief introduction to content strategy, including some of its important skills, genres of documents used in the field, and further readings to learn more
- Graduate students who are interested in learning specializations in content strategy, as well as more advanced skills, workflows, and best practices associated with content strategy
- Technical communication teachers looking to learn content strategy or to add it to their pedagogies within academic or workplace contexts
- Technical communication researchers seeking a reference guide as to best practices within this growing discipline
- Technical communication practitioners working in industry who are being called upon to do content strategy work in some way, shape, or form and need to quickly skill up in this field
- Technical communication practitioners seeking to inject content strategy best practices into their organizations through training or teaching

Defining Content Strategy

There are multiple definitions for content strategy that are currently used within the field. Rahel Bailie has defined content strategy as "a repeatable system that governs the management of content throughout the entire life-cycle" (as cited in O'Keefe & Pringle, 2012, p. 18). Additionally, Kristina Halvorson famously defined content strategy in a 2008 article as "planning for the creation, publication, and governance of useful, usable content." These two broad definitions still apply to a lot of what modern content strategists do:

- Planning: Content strategists are responsible for developing plans, templates, and guidelines for content.
- Creation: Content strategists are responsible for developing, or overseeing the development of, a wide variety of content, from reusable content stored in a repository to individual articles for a public-facing website.
- Publication: Content strategists are responsible for ensuring that content is published and delivered through the correct communication channels and is formatted correctly for specific genres that are appropriate to those channels.
- Governance: Content strategists are responsible for managing content once it's published, including keeping it updated, relevant, and authoritative over time, a process that should begin as soon as goals are set and technologies assembled to manage content.

If that sounds like a lot of work for one professional, it is! That's why content strategists often collaborate with other types of professionals who are responsible in some way for content. More than anything, however: content strategists *manage* content for their organizations. They create plans and strategies for the organization to follow and ensure everyone in the organization follows them.

At their heart, though, content strategists are effective writers and communicators. They have outstanding instincts when it comes to developing the right messaging for a specific audience. Whether they are creating content for a complex website or a simple handbook, content strategists understand how to craft a wide range of written genres for a wide range of audiences and how to manage this type of writing work in a way that melds well with organizational goals.

Tools of the Trade

Of course, lots of things have changed since Halvorson first brought content strategy to the attention of many content-focused professionals in 2008. Most notably, technologies have gotten more complex, including those that writers use to develop and deliver content.

Within your average organization, writers may be using technologies as diverse as:

- Desktop publishing software
- Collaboration tools such as video conferencing software and collaborative word processors (like Google Docs or OneDrive)
- Authoring tools that allow writers to quickly create content in a basic format and then output that content into a variety of other formats
- Open-source information architecture that helps writers structure content in such a way that it can be used by a variety of other people and technologies
- Content management systems (CMSs) or technologies that automatically store and format content for future use
- Component-based content management systems (CCMSs) that break content down into its basic components so that it can be reassembled later into complex genres such as large-scale manuals that need continuous updates and are distributed to thousands of audience members simultaneously
- Application programming interfaces (APIs) or tools within existing applications that enable writers to build their own, simpler applications, often for the purposes of storing content within those larger applications

Only the largest companies, such as Google, Apple, and Amazon, use many or all of these technologies for managing their content, but many companies use at least some of them. Such technologies help organizations, such as small, regional non-profits, mid-sized software development firms, and engineering firms, to develop, publish, and manage their content more efficiently than if they were writing without them.

Most notably, these technologies allow organizations to store content in a central repository that they can then draw on later for all the genres they need to write and deliver. This practice is commonly referred to as "single-sourcing" because the organization is able to keep all of their content in a single, authoritative repository that serves as the single source for all of their published genres.

Putting the Tools to Work: The Case of Educational Services

Just imagine for a moment being a technical writer for a regional educational services company that specializes in providing after-school programs and associated content to K-12 schools. These programs are provided through proprietary manuals on how schools can run each program, available for a subscription fee to the company. For an additional fee, staff of the company will administer the programs in-person, in which case the programs need to be reviewed by administrators of each school for compliance with their ongoing programs and policies. Because of these multiple audiences, each educational program manual has to be produced with very complex specifications in mind so everyone involved in their production, from school administrators looking for new after-school programming to teachers looking to bring new after-school programs to their school to parents trying to assess the appropriateness of programming for their individual child, need to have continual access to this educational manual content at all times. And if a policy or state law should change, this change needs to be instantly communicated to everyone involved in the process.

Before the internet, such technical content was much more difficult to manage. It had to be published in print and shipped to all the people associated with the educational services firm. For this to happen, it had to be written, edited for accuracy, formatted, printed, mass-produced, and shipped by snail mail. Then, if any aspect of the after-school programming changed, the process had to happen all over again. And some of these manuals were hundreds of pages long!

Now, imagine the same educational services firm today that has all of their content stored in a content management system. The content management system is set up to display the most current after-school programming for everyone involved. It is organized by the type of audience member trying to access it: administrator, teacher, parent, technical writer, or program staff. It is searchable by article, too, so people don't have to wade through hundreds of pages of writing to find the information they need. Most importantly, however, any changes can be made by logging into the system, making edits, and hitting publish.

If it is a very large company or has a smart content strategist, this content might also be hooked to a lot of other publishing platforms, such as the help forum for teachers looking to run their own after-school programs, a newsletter system for parents that provides updates about what their kids are doing in individual programs, and even a mobile app that allows administrators to look up the information they need about policy compliance on their smartphones. When a technical writer at the company updates any piece of content, it instantly updates to all these other genres as well.

Someone has to keep this system working, however. Someone has to oversee the whole process to ensure that the content is working properly, that it gets published to the correct formats, that it makes sense for each different audience it serves. Someone has to be responsible, in other words, for getting the right content to the right people at the right time for the right reasons.

That someone is a content strategist.

Introducing This Book

This book showcases many of the essential skills, workflows, and tools you'll need to become a content strategist.

Lots of books exist on what content strategy is and how to do it. What this book adds are:

- An easily understandable introduction to this emerging field catered toward students and others who are brand new to it
- A commonsense overview of the workflows and tools used by modern-day content strategists
- Examples of how to use these workflows and tools within actual projects
- Exercises and assignments educators and trainers can use to help any type of learner skill up in content strategy

In the past, it was enough to be a talented writer and editor to make it as a writing professional in a technical field, such as technical writing or technical editing. That simply isn't the case anymore. If you are someone who has a passion for delivering useful, usable information to people who need it, then you'll need to learn how to manage content and you'll need to master a variety of tools to do so.

This book has arisen from the experiences of the authors, some of whom have worked as active content strategists for several years now. All of us have discussed new developments in the field of technical communication with a host of practitioners, however. We've learned about the challenges they face on a daily basis. And we've learned how they solve problems.

Next we describe some of the ways these professionals solve problems.

The Strategy Part of Content Strategy

Content strategists do a lot of work to develop and publish useful, usable information for a variety of audiences. They typically do this by working within organizations to break down what Ann Rockley and Charles Cooper (2012) have called "content silos" (p. 133). Sticking with our running example, what if the education services firm, to save money, decided not to hire a content strategist. So, rather than building a content management system to manage all their content, they just have their technical writers work in isolation on individual pieces of content.

The writers then store these pieces of content wherever it is convenient for them to do so. Some versions are kept on the laptops of individual writers. Some get uploaded to a company intranet so they're available to other departments. These disparate content pieces then get used by marketers, customer service specialists, and other professionals within the company, and many of these professionals also create new versions of them that are in turn stored in other places.

Over time, the versions of these content pieces get stored up in these content silos. Because no one person is paying attention to all the content, it's getting created, and recreated, and reused willy nilly. School administrators are getting angry because the requirements they're asking customer service specialists and marketers for aren't ending up in the final after-school program manuals they receive. The company is losing money because of this as customers cancel their subscriptions and ask for a refund. This is further slowing down the process of producing program manuals for new customers, who are also getting angry about these delays.

Some of these customers decide to purchase their educational programs from another company because of this. The superintendent of one very large school district serving several dozen schools decides to sue the firm for breach of contract when they receive several manuals that aren't to spec. They want to be paid for the time lost, which has damaged their reputation with several parent associations and local school boards who have been waiting for the programs to be put in place and aren't seeing progress.

The educational services firm falls on hard times and eventually goes bankrupt, leaving hundreds of employees out of work.

This may seem like a doomsday scenario, but the current economy is littered with failed companies who were unable to adapt to current market demands. Many companies exist in very competitive marketplaces, meaning that if they fall behind, their competitors will begin to lure customers away.

Successful companies know that they need a strategy for their content that covers everything from marketing to customer acquisition to sales to customer service to customer support after purchase. Customers want to buy from highly organized companies that take care of them. They want to see the same information in every piece of content. If they see conflicting information, they lose confidence.

In order to ensure *all* the content an organization is publishing *to every source* is current, authoritative, and credible, someone has to manage that process. Someone has to make sure that new content is developed and published in tandem with organizational goals. Without that someone, or several someones if the organization is large or has very complex content needs, content will quickly become siloed again and will begin to drift away from organizational goals.

This is why many content strategists create frameworks, guidelines, and templates for content within their organizations. Known collectively as content models, these frameworks ensure that anyone developing content within the organization is following the same rules. These content models are often included as part of overall content strategy plans, which often include:

- Goals: measurable, achievable, simple, task-oriented objectives for what content should do for an audience
- Audiences: a descriptive rendering of each audience group as a persona or representative audience member
- Channels: a complete listing of all the various communication channels through which content will be delivered

- Content models: frameworks for messaging within each channel that align the specific requirements of the channel with content goals and audience needs
- Editorial calendars: lists of tasks that need to be accomplished on a regular basis, including which tools are best suited for which tasks

And, of course, as we've already mentioned, many of these content models are now built into the technologies that organizations use to write in the first place. That way the technologies control the formatting of every piece of writing produced and everyone is on the same page.

Regardless, someone has to make sure all of that happens. It's the content strategist's job to make sure that every audience the organization serves receives appropriate information.

Next, we turn to some of the ways content strategists ensure audiences are served.

Zeroing in on Audiences

Recall that the overall goal of content strategy is to plan for the creation, publication, and governance of useful, usable information. This also means doing so in a repeatable manner by creating consistently structured content for reuse, managing that content in a definitive source, and assembling content on demand to meet audience needs. All the best laid content plans will amount to nothing if they don't result in content that meets audience needs. Audiences drive this process. Content that is seen as useful by target audience members must be published in a timely manner and must reach each audience member in the appropriate genre—no small feat in the sea of communication channels that currently exist.

This is why content strategists need to pay close attention to audience needs. And they need to match those needs with appropriate communication channels and genres. A blog post probably isn't an appropriate genre for specific programming information tailored toward school administrators who are existing customers, for example. It is probably an appropriate genre for administrators who are shopping around for a new supplier of after-school programming, however, because it is easily findable in a search engine.

Such a blog post that is focused on customer acquisition might mention general aspects of after-school programs, such as the types of activities available, but wouldn't divulge such information as how to run entire programs for a given school year. Information like the latter should probably be part of a password-protected content management system that only existing customers with a paid subscription have access to. That way, the company can ensure that only paying customers have access to full programming and can indemnify themselves against anyone attempting to implement one of their after-school programs in a way in which it wasn't intended.

A content strategist serving these two very different audiences, new customers and existing customers, needs to produce very different types of content. In order to do so, they need to know their audience inside and out. They need to know what each type of audience expects, what they value, and what will most appeal to them. They need to know what type of information is useful and usable to each audience, in other words.

There are a lot of different research methods, or ways to gather data on audience preferences, such as:

- Surveys of audience members that ask them about demographic information and content preferences
- Interviews with randomly selected audience members to uncover more detailed needs and pain points
- Usability testing of content in a variety of media
- Analysis of data collected through analytics programs or through routine organizational processes such as consumer complaints or error reports
- Analysis of search engine data
- Analysis of usage data within a consumer-facing content management system or other platform

The goal of this audience analysis is to understand the needs of specific types of audience members in order to create content that appeals to each type of audience member who will see a specific piece of content.

Content strategists often display the results of audience analysis as personas, or archetypal audience members. Such personas typically include the following information for each type of audience:

- Name
- Photo
- Demographics (age, race, gender, location, occupation, etc.)
- Story: What makes them a good audience member for this content? What cultural values do they bring to the content?
- Goals and challenges: What is the audience member trying to accomplish with the help of this content? What pain points are they experiencing that can be alleviated through this useful information?
- How I can help: What can the strategist do through their content to help the audience member achieve their goals and alleviate their pain?

Many companies have multiple personas that they use across departments. These personas should be tied to actual customer demographics to ensure they represent real, live customers the organization is serving.

What the Future Holds for Content Strategy

Content strategy will only grow in importance as organizations, and the products and services they provide, increase in complexity. As this complexity increases, the content associated with it will also grow in complexity. All this complexity will require people who can make complexity simple for a specific audience.

Content will continue to be critical in all contexts, across all sectors of the world economy. The strategies for developing and delivering it to people will change, however. That's why the workflows presented in this book are technology- and genre-agnostic, meaning they can be applied in a variety of contexts, regardless of the tools available.

Book Takeaways

This is a book for content strategy students, teachers, researchers, and practitioners, as well as students within the field of technical communication and adjacent fields (user experience, marketing, communication, business, etc.).

In its pages you will learn:

- Best practices for the creation, publication, and governance of useful, usable information
- Best practices for managing content in a wide variety of formats
- Best practices for assembling content on demand to meet audience needs
- Best practices for creating repeatable systems for doing all these things

Overview of Book Chapters

Here is an overview of all the chapters in the book.

Chapter 1 introduces key concepts emerging within content strategy that represent the best practices that are central to the field. These concepts include intelligent content, unified content strategy, and the content strategy quad. The concepts are discussed in the context of a rapidly evolving field. The chapter ends with a list of further reading to help learners do their own research into content strategy.

Chapter 2 discusses many of the essential workflows of modern-day content strategists, including how to analyze an audience, how to audit existing content, how to model content, how to assemble a content strategy plan, how to work with other content developers, how to navigate content and constraints within an organization, and how to develop specific content genres for specific channels of communication while adhering to an organization's mission, vision, and goals. These workflows are explained as an overarching process of writing, editing, publishing, managing, and delivering technical content. Each succeeding chapter will then discuss each workflow in-depth, providing in-depth examples, discussion questions, example assignments, and learning outcomes for that specific workflow.

Chapter 3 delves into the intricacies of audience analysis, including identifying potential audiences, gathering audience data through different research methods, and developing audience personas. The overall goal of the chapter is to emphasize that a concrete sense of audience must be established before any other workflow can take place, and that depending on the audience, the type of content that needs to be created and accessed can vary.

Chapter 4 helps readers identify different content types and their importance to content strategy. In order to share content, you need to identify the different channels for delivering content, as well as how to develop a channel plan best suited to a specific situation. The chapter emphasizes that there is no one-size-fits-all approach to content development, but that content must be created with a specific audience and channel in mind.

Chapter 5 introduces readers to the essential task of content auditing, or the method of assessing existing content before developing new content. The chapter covers many aspects important to content auditing, including developing a content rubric, conducting a thorough content inventory, assessing content via a rubric, and reporting out findings to external audiences. This chapter also raises the importance of understanding the objective of your content audit, and auditing with a goal-oriented mindset. Additionally, this chapter discusses tools that are useful during this process, including site-mapping technologies and search engine web crawlers.

Chapter 6 delves into content modeling by providing a large list of genres that are common to content strategy in technical communication, including blog posts, social media posts, emails, webpages, and structured content built into a component-based content management system. Heuristics are also included in this chapter that will help readers quickly assess the key attributes of any genre, the appropriate genre for a specific channel, and how to produce content to those specifications.

Chapter 7 discusses the necessity of putting together a formal content strategy plan that contains goals, audiences, channels, content models for specific genres, and an ongoing editorial calendar that includes a workflow for making use of tools. This chapter helps readers develop such plans in a hands-on, contextualized manner, including by providing a planning cheat sheet developed by the authors.

Chapter 8 covers collaboration with other content developers, including technical writers, technical editors, subject matter experts, marketers, and managers. Specific workflows for collaboration that are common to content strategy are discussed, including organizing planning meetings, developing an editorial calendar, and revising and editing content.

Chapter 9 assists readers in revising and editing specific content genres through the introduction of best practices in peer review, editing, revision, and audience alignment.

Chapter 10 explains how learners can ensure their content is usable and accessible for their target audiences. The chapter includes explanations of usability testing research methods, accessibility guidelines, how to assess the usability and accessibility requirements of specific audiences, and how to adapt content to meet these requirements before delivery.

Chapter 11 covers the finer points of governing content throughout its lifecycle. Emphasis is placed on continuing the best practices introduced throughout the book including audience analysis, content auditing, and content strategy planning to ensure content remains relevant, authoritative, and credible over time.

Chapter 12 explores how to localize content to specific cultures, including international and non-English-speaking audiences. Emphasis is placed on the differences between localization, which focuses on cultural knowledge, and translation, which focuses on linguistic variance. Transcreation of genres across cultures is also discussed, as are strategies for localizing multilingual content within disparate cultures while it is being developed.

Chapter 13 introduces learners to some of the core tools and technologies of content strategy, including desktop publishing, collaboration tools, authoring tools, content management systems (CMSs), component-based content management systems (CCMSs), and application programming interfaces (APIs). Using Hovde & Renguette's (2017) technological literacy framework, each technology is explored within the context of a specific type of content project so that readers can evaluate its function and use within a specific context, as well as its place within the broader content strategy landscape.

Finally, Chapter 14 concludes the book by discussing strategies for formal education, lifelong learning, and career development in content strategy. The chapter includes exercises learners can engage in to help them network with others interested in content strategy, to develop an active portfolio in the discipline, and to seek out related internships and jobs.

Getting Started Guide: Exploring Content Strategy

To explore the burgeoning world of content strategy, try doing some of your own research about the field. Review one or more of the sources listed in the Further Reading section. Also, try using Google (or your other favorite search engine) to search for phrases like:

- What is content strategy?
- Content strategy definition
- Content strategy skill sets

Peruse the information you find and use it to answer some of the following questions about the topics presented in this chapter:

1. What alternative definitions of content strategy did you come across? How do these alternative definitions compare with the definitions presented in this chapter?
2. What specific skill sets did you see mentioned in blogs and articles about content strategy? How do these skill sets compare to the ones presented in this chapter?

References

Halvorson, K. (2008). *The discipline of content strategy.* A List Apart. Retrieved from: http://alistapart.com/article/thedisciplineofcontentstrategy.

Hovde, M. & Renguette, C. (2017). Technological literacy: A framework for teaching technical communication software tools. *Technical Communication Quarterly*, 26(4), 395–411.

O'Keefe, S. S. & Pringle, A. S. (2012). *Content strategy 101: Transform technical content into a business asset.* Scriptorium Publishing Services, Inc.

Rockley, A. & Cooper, C. (2012). *Managing Enterprise Content: A Unified Content Strategy.* 2nd ed. New Riders.

Further Reading

Andersen, R. (2015). The emergence of content strategy work and recommended resources. *Communication Design Quarterly*, 2(4), 6–13.

Content strategy: The SEO's guide to content marketing. Moz. (n.d.). Retrieved January 20, 2022, from https://moz.com/beginners-guide-to-content-marketing/content-strategy

Department of Health and Human Services. (2016, January 24). *Content strategy basics.* Usability.gov. Retrieved January 20, 2022, from www.usability.gov/what-and-why/content-strategy.html

Halvorson, K. (2017, October 26). *What is content strategy? Connecting the dots between disciplines.* Brain Traffic. Retrieved January 20, 2022, from www.braintraffic.com/insights/what-is-content-strategy

Patel, N. (2022, January 7). *Content strategy: What is it & how to develop one [2022].* Neil Patel. Retrieved January 20, 2022, from https://neilpatel.com/blog/content-strategy-a-development-guide/

1 Key Concepts in Content Strategy

As a field, content strategy has been around for some time now. Kristina Halvorson notably defined content strategy in a 2008 article as "planning for the creation, publication, and governance of useful, usable content," so many mark that year as the birth date of the field. However, strategists like Halvorson were already working to develop content in a strategic manner, mostly for websites, for several years before she wrote that article. Halvorson's article and her ensuing book *Content Strategy for the Web* certainly put a name to this burgeoning collection of professional practices, as well as bringing it into the professional spotlight as a field in its own right.

As McCoy (2021) pointed out, the *practice* of content strategy dates back to the advent of early publications like *Poor Richard's Almanack*, a genre that Benjamin Franklin used to disseminate aphorisms and proverbs, many of which are still with us to this day. Franklin, being the savvy content creator he was, knew that simply disseminating practical information on the calendar, weather, astronomy, and astrology of his time wasn't enough to capture an audience's attention. So, Franklin created the persona of Poor Richard, a relatable astrologer and lover of learning who told witty stories that readers of the time could relate to. These stories were serialized, meaning that people had to buy the newest edition to read the next installment.

This was content strategy, because the *content*, the *information used by audiences*, was the selling point of Franklin's publication, not the container, or *channel*. There were a number of almanacs being published at the time. Franklin's version sold so well because he gave his audience something extra, something they couldn't get from other publications, something they could relate to as people.

At the same time, his almanac made use of cutting-edge technology, at least for the mid-eighteenth century. Though the printing press had been invented centuries before, there were few citizens of the original 13 colonies that would become the United States who had access to one, much less the ability to utilize one for their own economic benefit. The ability to publish

DOI: 10.4324/9781003164807-1

and distribute high-quality, printed content was the best way to reach a large audience during this time.

Hundreds of years later, content strategists are still attempting to do what Franklin did so well: to create timely content that will reach and engage its target audience. Things are a *little* more complex now, however. We have internet, mobile devices, wearable devices, e-readers, and thousands of applications that run on these technologies. We have websites, mobile apps, augmented reality, and even virtual reality.

The challenge of specializing in a field like content strategy is that the means of distributing content change constantly. At the same time, there are technologies that have been with us since the early days of the internet, such as search engines and webpages. Print is very much still a thing, with book sales still regularly exceeding 650 million units per year (Watson, 2021). Rather than subtracting technologies since print was the only real means of disseminating information to a broad audience, we've added many new technologies that people can interact with in completely novel ways.

Social media was just getting started when Halvorson first declared content strategy a profession in 2008, as one example. Now we have dozens of social media applications, with new ones being rolled out every year. As a case in point, only a handful of social media channels have emerged that capture a sizable portion of internet users. Many content channels don't make the cut and end up falling by the wayside. Some readers of this book may remember a little website called MySpace that captured the collective imagination of an entire generation. And though it continues to exist in a new form (*https://myspace.com/*), it is no longer the preeminent virtual meeting space it once was.

What all of these technologies have in common is that they all:

- Are centered around content, or useful information
- Provide that content to a defined audience of people
- Are used by organizations to promote products and services
- Use some kind of publishing model that distributes content in a definable manner
- Are businesses in their own right, meaning they profit off of the distribution of content, typically through selling advertisers the ability to reach audience members

Outside of these broad strokes, however, all content-based technologies are very different from one another. They use different means of disseminating content. They target different groups of people. They allow people to use, distribute, and create their own content in different ways. And they use different content types to accomplish all this.

These differences are the key to concepts we've seen arising in content strategy since 2008. At time of writing, no one person can claim to be a strategist of every type of content that exists. No one person can claim to be an expert on distributing content via all the currently available channels. Rather, today's content strategists often brand themselves as experts of particular types of content, particular types of channels, even particular industries.

Employers want content strategists with experience that closely mirrors their organizational goals. And with organizations ranging from K–12 schools to large technology companies discovering and utilizing content strategy, this means there are many, many opportunities for specialization in this field.

As mentioned earlier, specialization can also mean one's approach to particular technologies, content types, or content channels. Or it can mean a specific approach to *the very practice of content strategy itself* that sets one apart from what other practitioners are doing. Most people define this kind of specialization in a field as a "best practice," meaning a practice that other members of the field should follow. We'll start with this broadest definition of specialization here because we think it will help readers best understand the current shape of this field.

Content Strategy Best Practices

As a rule, we're going to introduce content strategy concepts in this section that appear to fall under the heading of "best practices," meaning that a lot of people in the field of content strategy continue to agree that they're useful in many situations. Many best practices have fallen by the wayside over the years as technologies change or audience needs shift. Older readers of this book may remember something called RSS (or Really Simple Syndication) that was once an effective way for website owners to allow users to subscribe to their website for updates. As open-source content management systems (CMSs) like WordPress, Drupal, and Joomla! began to develop their own internal features for subscribing users that came pre-packaged with their default settings, RSS slowly lost its dominance as a delivery mechanism for website content (though podcasters still use it).

There is also a lesson about content strategy best practices in this story, however, because from our viewpoint as people who have followed the content strategy conversation for many years now, we've seen many practices come and go. The ones that seem to endure are mostly technology-agnostic, meaning they focus on the *strategy* part of content strategy, rather than on a very specific content type or content channel.

One of the first such concepts was Kristina Halvorson's Content Strategy Quad (Figure 1.1):

Figure 1.1 The content strategy quad

As Halvorson's company Brain Traffic describes the quad, "At the center is the core content strategy, the approach you'll take with your website, product, or service content to meet user needs and achieve your business goals" (Brain Traffic, 2017). As you can see from the image, this core content strategy involves separating content components from people components. First published in her 2012 book *Content Strategy for the Web* with Melissa Rach, the quad was one of the first depictions of content strategy as involving both content and people. It was also one of the first depictions to break up content strategy into four interlocking concerns: substance, structure, workflow, and governance. Put simply:

Content-focused components

• Substance: What kind of content do we need (topics, types, sources, etc.), and what messages does content need to communicate to our audience?
• Structure: How is content prioritized, organized, formatted, and displayed? (Structure can include communication planning, IA, metadata, data modeling, and linking strategies)

People-focused components

• Workflow: What processes, tools, and human resources are required for content initiatives to launch successfully and maintain ongoing quality?
• Governance: How are key decisions about content and content strategy made? How are changes initiated and communicated?

(Brain Traffic, 2017)

The impact of these ideas on the field of content strategy can't be overstated. The quad was arguably the first time anyone had created a simple infographic that depicted what content strategy is. And reading through the dozens of books and articles that have been published in the field since, the idea that content strategy involves both people and content components that are mobilized in a variety of strategic processes is still at the heart of how content strategy is thought of today.

Another early concept for the entire process of content strategy is the Unified Content Strategy. First introduced by Rockley & Cooper (2012), they defined this concept in the following way:

A unified content strategy is a repeatable method of identifying all content requirements up front, creating consistently structured content for reuse, managing that content in a definitive source, and assembling content on demand to meet customer needs. A unified content strategy can help organizations avoid the content silo trap, reducing the costs of creating, managing, and distributing content, and ensuring that content effectively supports both organizational and customer needs.

(p. 10)

Some keywords that would heavily influence later thoughts on content strategy here are "repeatable," "structured," and "silo." Essentially, Rockley and Cooper were arguably the first thinkers to identify a major problem within many organizations: that they are creating content in a haphazard manner without a repeatable system in place. This siloing of content causes a lot of waste as individual content creators between departments in the same organization duplicate efforts, or even create content that works at cross purposes against the content of other departments.

Another cause of content siloing is the creation of content in a specific format. If you publish content in a webpage and then try to use that same content in a help doc, often the formatting doesn't translate between channels. This creates further waste as you have to reformat content constantly. Rockley and Cooper were early adopters of the idea of "structured content," meaning content that is stripped of a specific generic format (e.g., blog post) and stored in its most basic form (e.g., as textual information in a database). We discuss structured content more thoroughly in Chapter 8, but the key point here is that structuring content allows content creators to pull up content in its basic form and then publish it on demand in specific formats without having to remove all the markers of its last published format.

An important attribute of content that content strategists should strive for that was also introduced by Rockley and Cooper is the idea of intelligent content. As they put it, "[i]ntelligent content is designed to be modular, structured, reusable, format free, and semantically rich and, as a consequence, discoverable, reconfigurable, and adaptable" (p. 16). As they would later clarify with their coauthor Scott Able in the book *Intelligent Content: A Primer.* "There are two parts to this definition. The characteristics that make content intelligent (modular, structured, reusable, format free, and semantically rich) and the capabilities we gain from adding intelligence to our content (discoverability, reconfigurability, and adaptability)" (Rockley, et al, 2015, p. 1). Essentially, intelligent content is differentiated from regular content because it's reusable and can be easily adapted to new situations, largely through stripping it of its generic format as we described earlier.

As we've mentioned before, a lot of what has driven the evolution of content strategy are changes in technology. Changes in technology mean that organizations often have to create content differently than they did before. Thus, in 2018, Halvorson revamped the Content Strategy Quad to better account for modern processes of content development (Figure 1.2).

This new version of the quad attempted to account for the growing complexity of content development in a world with an increasing array of devices and user needs. As Halvorson (2018) herself put it: "it's next to impossible to separate out workflow from governance—one can't (or shouldn't) exist without the other." This redefined quad also introduced two new keywords that are growing in importance as content strategy evolves: content design and systems design.

According to Halvorson, content design, a term she borrows from Content Design London (n.d.), is "the process of using data and evidence to give the audience the content they need, at the time they need it, and in a way they expect." So, this definition of content creation shows the shift in thinking toward several aspects that are increasingly important for content strategy: audience targeting, timeliness, and the role of technology in delivering content. As we write this book, there are hundreds of different types of devices available to consumers, and almost all of those devices serve consumers some type of content. And in the coming years, even

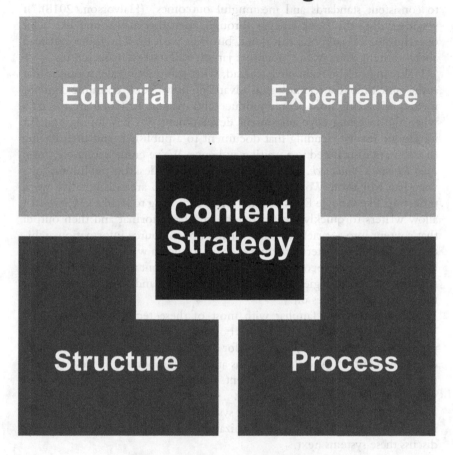

Content design

Editorial

Experience

Content Strategy

Structure

Process

Systems design

Figure 1.2 The updated content strategy quad

more types of devices will be invented and made available to the wider public. This is why *design* is now a key factor in how effective content is: content has to interact with a specific device and the content genres that device supports.

Systems design, on the other hand, is "the process of defining the architecture, modules, interfaces, and data for a system to satisfy specified requirements." She goes on to explain: "[w]e're interested in creating repeatable systems—both for machines and for people—to ensure content integrity over time and allow us to create, deliver, and manage content according to consistent standards and meaningful outcomes" (Halvorson, 2018). If you're reading this book straight through, then this may sound like a lot of technological jargon at this point, but bear with us. The more you read about content strategy, the more that jargon will start to make sense!

Following on the technology thread, what this quote really means is that at the same time that technology is advancing how people consume content, it's also advancing how people produce and manage it. Gone are the days when the bleeding edge of content development was developing content in a single format, sending that document to a publisher, and then having that content distributed to a waiting audience. We've come a long way from *Poor Richard's Almanack*. And don't get us wrong, desktop publishing software like Microsoft Word is still important. But so are collaborative word processors like Google Docs or OneDrive, authoring tools like Oxygen that allow writers to quickly create content in a basic format and then output that content into a variety of other formats, open-source information architecture that helps writers structure content in such a way that it can be used by a variety of other people and technologies, content management systems (CMSs), or technologies that automatically store and format content for future use, the list goes on.

And if you're not familiar with most of these technologies, that's ok! You'll learn about them later in this book, specifically in Chapter 13. The point is: there are entire *systems* now for developing, storing, publishing, and delivering content. And these systems are just as important as the technologies that allow people to access content. You have to take them into account when you develop a content strategy.

And there are two very important systems, websites and internal content repositories, that almost every organization has to contend with now. We discuss these systems next.

Websites and Internal Content Repositories

The rise of content strategy as a field can be explained as the rise of many fields probably can: as a response to specific types of problems. To illustrate two very common problems content strategists face, let's look at two quotes from some of the thought leaders in the field.

First, this is how Halvorson & Rach (2012) explain why content strategy became important for websites:

> While organizations have struggled for decades—centuries, even—to make sense of their content, they were always able to keep the chaos (and consequences) to themselves. Then came websites, which created the perfect content strategy storm. Suddenly, organizations had to put

all of their content (product info, investor reports, press releases, etc., etc.) in one place. For the first time. For all the world to see. And it hurt.

(p. xvii)

If you're below the age of 30, you might not remember how much the information we use on a daily basis changed, seemingly overnight, with the advent of websites. Some of us can remember having to call businesses on the phone (or even send them a letter!) to request service. Suddenly, tons of information about everything you could possibly imagine was available online. Many organizations struggled, and continue to struggle, with being always available to consumers.

And websites have only grown in complexity over the years. Now they must attract users via search engines to stay relevant. They must display well on mobile devices. They must utilize a variety of features, from automated newsletters to GPS-enabled advertising. The requirements for a modern website are only growing by the day. Imagine being a content strategist for a large organization when all you had to do was turn information into webpages! Now you might need to manage such diverse channels as webpages, blogs, social media, email newsletters, white papers, journal articles, paid online advertisements, print catalogs, books, ebooks, the list goes on!

If you're hosting a website of a certain level of complexity, working with a content strategist is now a necessity, unless, of course, you like to struggle and experience lots of problems. And believe it or not, many managers and executives continue to choose to struggle rather than hiring a competent content strategist. In our personal experience, in fact, this still happens more often than not. But at the same time, many, many people who run organizations, be they universities, hospitals, or engineering firms, are learning how important content strategy is to ensure they have an effective web presence. And that's one reason why we're seeing such job growth in the field.

Another reason we're seeing growth in content strategy jobs, however, is what happens *within* organizations. To explain what happens with content creation within all too many organizations, we turn again to Rockley & Cooper (2012) who describe something they call the "content silo trap":

> Too often, content is created by authors working in isolation from others within the organization. Walls are being erected between content areas and even within content areas. This leads to content being created, and recreated, and recreated, often with changes or differences introduced at each iteration. No one has a complete picture of the customer's content requirements and no one has the responsibility to manage the customer experience.

(p. 5)

As messy as many organizations are with their web presence, inside they are even worse. Just imagine the mayhem that an organization with thousands of employees can get up to when creating content for no rhyme or reason.

Collectively, we've spoken to hundreds of content creators who describe conflicts between individuals, departments, goals, and others within their workplaces. We've spoken to technical writers who are practically at war with their marketing department who make promises to customers that the organization's product or service simply can't fulfill. We've spoken to marketers who beg technical writers to allow them to use information about a product or service that is getting lots of hits over search engines, but the technical writer sees the information as "theirs" and won't allow it to be listed on the website. We've spoken to managers who put their content into a particular format years ago (such as PDF) that they were told was "universal" and now they need to put it into a format that didn't exist when it was first created (such as an internal content management system that also publishes content to the organization's website), but they can't easily free the content from the format it was created in without reproducing it all from scratch.

And we've spoken to many, many professionals who are responsible for content and who simply don't know what to do with it. Should they put it into a content management system like WordPress? Should they store it on a hard drive? Should they put into the cloud? Should they buy software to manage it? Should they do all of these things? As technologies for creating, storing, publishing, delivering, and managing content have multiplied, so have the options for content creators. Just try doing a Google Image Search for "content flow" to see all the depictions experts have created for the ways content goes through an organization to its intended audience.

And this has led many content strategists to the startling conclusion: maybe there isn't *one right way* to do content strategy for every organization out there. Like the field of technical communication, we're starting to see the rise of *subject matter expertise* as an important factor for content strategists. Members of the field are learning that it's going to be different working as a content strategist for a software development company than it is working for a K–12 school.

Of course, not every industry out there is hiring content strategists at the same rate. There are far more content strategy jobs in the technology sector of the economy (i.e., electronics, software, computers, artificial intelligence) than in other sectors, such as education. As all of the authors of this book work at universities or have worked at universities, we are *painfully* aware of how poorly represented content strategy is in higher education, for example. Many colleges and universities simply purchase a content management system for their website and call it good. Many don't even have a central place to store content, allowing it to be controlled by dozens of different technologies. And thus, many of these organizations are wasting precious resources every year creating and recreating (and recreating!) content that could be better managed.

So what industries are hiring the most content strategists? Well, we simply don't have comprehensive data on that question right now. Content strategy is so new compared to a field like marketing that there simply hasn't been a truly collective effort by the field to track its job growth, though several thought leaders have conducted surveys with small samples of people (i.e., *The state of content*, n.d.). As people who have been paying attention to this field for many years now, however, we do have some resources to share with you where you can join the conversation.

We turn to those resources next.

Sources of Information on Content Strategy (AKA Further Reading)

What we can't present to you in this section, for the reasons we described earlier, is a complete breakdown of which industries are hiring content strategists and at what rate. There simply isn't a centralized database of information out there that we could analyze to tell you that. What we *can* do is tell you where you can find the highest concentration of content strategy jobs and up-to-date information on the field so that you can investigate this question for yourself.

This will also not be a comprehensive list of all the sources of information on content strategy out there. That would be a book in and of itself and we want this book to be an introduction to how to *do* content strategy. This list is strongly shaped by our own experience with content strategy. In other words, by presenting this list we're saying that we've personally found all of these sources to be useful. And we're also saying that if you peruse these sources of information, which we break down by genre next, you will gain a much better understanding of the state of the field than any one book can provide. We should also note that this list of resources will take the place of the Further Reading section you'll see in every other chapter of this book. We thought this was necessary because it's such a long list of sources and because every other Further Reading section is a lot more tailored to the topics presented in each specific chapter.

Following this list, we close this chapter with our thoughts on the future of content strategy.

Without further ado, here's our semi-comprehensive list of sources on content strategy!

Academic Journal Articles and Special Issues

Bailie, R. A. (Ed.). (2019). Special issue on content strategy. *Technical Communication*, 66(2), 121–199.

Batova, T. (2018). Negotiating multilingual quality in component content-management environments. *IEEE Transactions on Professional Communication*, 61(1), 77–100.

Batova, T. & Andersen, R. (Eds.). (2015). Special issue on internal content management. *Transactions on Professional Communication*, 58(3), 241–347.

Batova, T. & Andersen, R. (Eds.). (2016). Special issue on web-based content strategy. *Transactions on Professional Communication*, 59(1), 1–67.

Pullman, G. & Gu, B. (Eds.). (2008). Special issue on content management. *Technical Communication Quarterly*, 17(1), 1–148.

Walwelma, J., Sarat-St. Peter, H., & Chong, F. (Eds.). (2019). Special issue on user-generated content. *IEEE Transactions on Professional Communication*, 62(4), 315–407.

Articles Published to the Web

Department of Health and Human Services. (2016, January 24). *Content strategy basics*. Usability.gov. Retrieved September 16, 2021, from www.usability.gov/what-and-why/content-strategy.html

Halvorson, K. (2008). *The discipline of content strategy*. A List Apart. Retrieved January 20, 2022 from: http://alistapart.com/article/thedisciplineofcontentstrategy

Books by Academics

Albers, M. & Mazur, M. (Eds.). (2003). *Content and complexity: Information design in technical communication*. Routledge.

Bridgeford, T. (Ed.). (2020). *Teaching content management in technical and professional communication*. Routledge.

Getto, G., Labriola, J. T., & Ruszkiewicz, S. (Eds.). (2019). *Content strategy in technical communication*. New York, NY: Routledge.

Books by Practitioners

Abel, S. & Baile, R. A. (2014). *The language of content strategy*. XML Press.

Atherton, A. & Hane, C. (2017). *Designing connected content: Plan and model digital products for today and tomorrow*. New Riders.

Baile, R. A. & Urbina, N. (2012). *Content strategy: Connecting the dots between business, brand, and benefits*. XML Press.

Bloomstein, M. (2013). *Content strategy at work: Real world stories to strengthen every interactive project*. New York, NY: Morgan Kaufmann.

Casey, M. (2015). *The content strategy toolkit: Methods, guidelines, and templates for getting content right*. New Riders.

Frick, T. & Eyler-Werve, K. (2015). *Return on engagement: Content strategy and web design techniques for digital marketing*. 2nd ed. Taylor & Francis.

Halvorson, K. & Rach, M. (2012). *Content strategy for the web*. 2nd ed. New Riders.

Land, P. (2014). *Content audits and inventories: A handbook*. XML Press.

McCoy, J. (2017). *Practical content strategy & marketing: The content strategy & marketing course guidebook.* CreateSpace Independent Publishing Platform.

Metts, M. & Welfle, A. (2020). *Writing is designing: Words and the user experience.* Rosenfeld Media.

Nichols, K. & Rockley, A. (2015). *Enterprise content strategy: A project guide.* XML Press.

Podmajersky, T. (2019). *Strategic writing for UX: Drive engagement, conversion, and retention with every word.* O'Reilly Media.

Reddish, J. (2012). *Letting go of the words: Writing web content that works.* 2nd ed. Morgan Kaufmann.

Richards, R. (2017). *Content design.* Content Design London.

Rockley, A. & Cooper, C. (2012). *Managing enterprise content: A unified content strategy.* 2nd ed. New Riders.

Rockley, A., Cooper, C., & Abel, S. (2015). *Intelligent content: A primer.* XML Press.

Wachter-Boettcher, S. (2012). *Content everywhere: Strategy and structure for future-ready content.* Brooklyn, NY: Rosenfeld Media.

Welchman, L. (2015). *Managing chaos: Digital governance by design.* Brooklyn, NY: Rosenfeld Media.

Wilson, P. (2018). *Master content strategy: How to maximize your reach and boost your bottom line every time you hit publish.* BIG Brand Books.

Blogs

Blog. Content Company, Inc.—Digital & Content Strategy Consulting for Content-Rich Websites—Hilary Marsh. Retrieved January 20, 2022, from https://contentcompany.biz/blog/

Blog. Content Garden. (n.d.). Retrieved January 20, 2022, from www.contentgarden.org/blog/

Blog. Scriptorium. (n.d.). Retrieved January 20, 2022, from www.scriptorium.com/blog/

Blog. The Content Wrangler. Retrieved January 20, 2022, from https://thecontentwrangler.com/blog/

Content strategy articles. Brain Traffic. Retrieved January 20, 2022, from www.braintraffic.com/insights/

Conferences and Professional Societies

Confab: The content strategy conference. (n.d.). Retrieved February 17, 2022, from www.confabevents.com/

Technical Communication Summit. (2021, December 6). Retrieved February 17, 2022, from https://summit.stc.org/

The LavaCon Content Strategy Conference. (n.d.). Retrieved February 17, 2022, from https://lavacon.org/

Society for Technical Communication. (n.d.). Retrieved February 17, 2022, from www.stc.org/

Job Search Engines

Indeed. (n.d.). Retrieved January 20, 2022, from www.indeed.com/

LinkedIn job search: Find us jobs, internships, jobs near me. LinkedIn. (n.d.). Retrieved January 20, 2022, from www.linkedin.com/jobs/

Salary database. Society for Technical Communication. (n.d.). Retrieved January 20, 2022, from www.stc.org/publications/salary-database/ (Note: requires membership to view)

Social Media

Content strategists. Facebook. (n.d.). Retrieved January 20, 2022, from www.facebook.com/groups/132535916799137

Content strategy. LinkedIn. (n.d.). Retrieved January 20, 2022, from www.linkedin.com/groups/1879338/

Welcome to Content Strategy. (n.d.). Retrieved January 20, 2022, from https://community.content-strategy.com/

University Programs

Content Strategy Certificate Program. Content Strategy Certificate Program | Northwestern SPS: School of Professional Studies. (n.d.). Retrieved March 21, 2022, from https://sps.northwestern.edu/advanced-graduate-certificate/content-strategy/#Content%20Strategy%20Required%20Courses

FH Joanneum Content Strategy. Content-Strategie/Content Strategy. (n.d.). Retrieved January 21, 2022, from www.fh-joanneum.at/content-strategie-und-digitale-kommunikation/master/en/programme/

Transmedia Certificate Program: Content Strategy. Content Strategy—University of Houston. (2022, February 15). Retrieved March 21, 2022, from https://uh.edu/tech/digitalmedia/transmedia/content_strategy/

UW Professional & Continuing Education. (n.d.). *Certificate in storytelling & content strategy.* UW Professional & Continuing Education. Retrieved March 22, 2022, from www.pce.uw.edu/certificates/storytelling-and-content-strategy

Industry Certificates

Content Strategy Course Inc. (2022, March 3). *The content strategy & marketing course: Your anchor in the Content Sea.* Content Strategy & Marketing Course. Retrieved March 22, 2022, from https://contentstrategycourse.com/

Content strategy course: Learn how to create a successful content strategy. HubSpot Academy. (n.d.). Retrieved March 22, 2022, from https://academy.hubspot.com/courses/content-strategy

Content Strategy in the Future

This final section of this chapter is where we really go out on a limb, because predicting the future of a field as rapidly changing as content strategy is like trying to predict the stock market. The easy answer is that we think the future continues to be very bright for this field. There continues to be a lot of demand for the skills associated with content strategy and too few people to fill these roles. This is partially because higher education hasn't kept up with content strategy like it has with other emerging fields, such as UX. There are individual courses in content strategy springing up in technical communication, communication, and marketing programs, but there are precious few comprehensive programs in it like there are with UX (we only know of one at a school in Germany, currently: FH Joanneum, n.d.).

As far as future trends in the field, we predict that, like technical communication and other content-focused fields, subject matter expertise will continue to matter. Someone with insider experience as a content strategist for healthcare will probably be better positioned within that industry than someone with experience as a content strategist for software. Beyond subject matter expertise, however, technical fluency will continue to be very important. This means that if an organization needs help taming their website that is built in WordPress, familiarity with WordPress, or at least familiarity with open-source content management systems, will be important.

At the same time, the non-technical skill sets associated with content strategy that we've mentioned so far, and will continue to explore throughout this book, continue to be the real heart of this field. We explore these skill sets in-depth in Chapter 2 and so won't go into them here, but they include skills like audience analysis, content auditing, and content governance. These will continue to be the threshold skills of content strategy, meaning the ones that will differentiate someone who has real expertise in content strategy from someone with a passing interest.

And we may see increased specialization in these individual skills, like we're seeing in related fields. There may be jobs where you just do content auditing or just do audience analysis in the near future. Right now, it seems like most of the jobs we see for content strategists want all the skill sets they can pack into an individual. They want someone to run the whole content process, rather than taking on a piece of it. That being said, there are related roles like "content manager," "website manager," and "content creator" you may see in job search engines. These are certainly in the vicinity of content strategy and are likely a great entry point into the field.

And there's always the omnipresent role of the content strategy consultant. Many of the thought leaders we've mentioned in this chapter—Kristina Halvorson, Rahel Anne Baile, Scott Abel, Ann Rockley, and the like—are folks that get paid to be themselves. They work outside of any one organization by consulting with lots of different organizations to help them improve their content strategies. They are the godmothers and godfathers of this field and will largely continue to shape it in years to come.

Outside of these few predictions? We don't have a lot for you, unfortunately. We're just beginning to see the impact of something like artificial intelligence (AI) on content strategy with the advent of things like chatbots within various forms of content (e.g., user help forums). We don't really know yet how something like the Internet of Things will affect how people consume content, because the technology that powers IoT isn't widely available to most consumers.

One thing is certain: there will be a continued need for people who can help meld organizational goals and audience goals. Organizations will continue to need people who understand why human beings consume content in the ways that they do. Many of the tools, skills, and exercises throughout the rest of this book were developed by just such people. We welcome you to this rich, vast field filled with so much potential. Happy exploring!

Getting Started Guide: Exploring Key Concepts in Content Strategy

To explore key concepts in content strategy, try doing some of your own research about the field. Review one or more of the sources listed in the Sources of Information on Content Strategy (AKA Further Reading) section. Also, try using Indeed (or your other favorite search engine for jobs) to search for jobs in content strategy. Current jobs being advertised are probably the best source to see what skills are in demand right now.

Peruse the information you find across 3–5 job ads that peak your interest and use them to answer some of the following questions about the topics presented in this chapter:

1. What are the primary skill sets the employer is looking for?
2. Do these skill sets use any of the keywords introduced in this chapter (i.e., intelligent content, content design, systems design)?
3. Does the employer want a content strategist who is familiar with a particular industry context (i.e., medicine, education, engineering, software)?
4. What "soft skill" attributes does the employer want in a content strategist (i.e., ability to work in cross-functional teams, ability to work with a specific other type of professional such as a software designer, ability to work with subject matter experts)?

References

Brain traffic lands the quad! Brain Traffic. (2017, July 6). Retrieved January 7, 2022, from www.braintraffic.com/insights/brain-traffic-lands-the-quad

FH Joanneum Content Strategy. Content-Strategie/Content Strategy. (n.d.). Retrieved January 21, 2022, from www.fh-joanneum.at/content-strategie-und-digitale-kommunikation/master/en/programme/

Halvorson, K. (2008). *The discipline of content strategy*. A List Apart. Retrieved from: http://alistapart.com/article/thedisciplineofcontentstrategy

Halvorson, K. (2018, April 26). *New thinking: Brain traffic's content strategy quad.* Drain Traffic. Retrieved January 7, 2022, from www.braintraffic.com/insights/new-thinking-brain-traffics-content-strategy-quad

Halvorson, K. & Rach, M. (2012). *Content strategy for the web.* 2nd ed. New Riders.

Homepage. Content Design London. (n.d.). Retrieved January 20, 2022, from https.//contentdesign.london/

McCoy, J. (2021, April 8). *The history of content strategy: A 10-year-old industry (infographic).* Content Strategy & Marketing Course | Proven Training Course from the Content Hacker™. Retrieved January 6, 2022, from https://contentstrategycourse.com/history-of-content-strategy/

Rockley, A. & Cooper, C. (2012). *Managing enterprise content: A unified content strategy.* 2nd ed. New Riders.

Rockley, A., Cooper, C., & Abel, S. (2015). *Intelligent content: A primer.* XML Press.

The state of content strategy 2021 report. Kontent by Kentico. (n.d.). Retrieved January 21, 2022, from https://kontent.ai/resources/state-of-content-strategy-2021-report/

Watson, A. (2021, September 10). *U.S. book industry—statistics & facts.* Statista. Retrieved January 6, 2022, from www.statista.com/topics/1177/book-market/#dossierKeyfigures

2 The Content Strategy Process

The Content Strategy Process: A Brief Overview

So, now that we've introduced what content strategy is and discussed some specializations within it, it's time to talk about *how to do content strategy*. Like most communication-centered activities, content strategy involves a lot of different processes geared toward creating the right content for the right people for the right reasons at the right time.

Content strategy involves so many different processes, in fact, that the rest of this book is devoted to discussing each one of them in-depth. This is not a comprehensive book. There are just too many different people doing content strategy in too many different ways to represent that in a single volume.

Rather, this is a breakdown of many of the processes that content strategists engage in. It is a list of best practices, practices that many (if not most) content strategists would agree on.

First, let's list the parts of the content strategy process that will be discussed throughout the remaining chapters of this book:

- Audience analysis
- Identifying content types and channels
- Content auditing
- Assembling a content strategy plan
- Collaborating with other content developers
- Content modeling
- Revising and editing genres
- Ensuring content usability and accessibility
- Delivering, governing, and maintaining genres
- Localizing content
- Using tools and technologies

Next, let's do a brief overview of the entire content strategy process and how it works. Then, we'll talk about the purpose of each process used by content strategists.

DOI: 10.4324/9781003164807-2

Creating the Right Content for the Right People at the Right Time for the Right Reasons

If you need a single sentence to carry with you as you journey through this book and learn about content strategy, it's this one: *the purpose of content strategy is to create the right content for the right people for the right reasons at the right time.* The overall goal of content strategy is to give audiences the best content you can deliver to them. Savvy content strategists learn quickly how to understand the psychology of their audiences and how to deliver content that meets their needs.

So, let's break this statement down further into its components.

Creating the Right Content

First, it's important for a content strategist to assess what content is the right content to create and deliver. Within any industry, there are lots of different types of content. In healthcare, for example, there are highly protected forms of content, like patient files, and then there are public forms of content, like common diagnoses. Then there are lots of different types of content in-between. There are websites that cater to people who want to Google their symptoms in order to see if their condition is serious or not.

There is also specialist content that you literally need a medical degree to understand. And there is technical content involving the tools, technologies, and data that healthcare providers and insurers use to make decisions, like what care to provide a particular patient, how much it will cost, and what insurance will cover. There are compliance reports that providers must write and file with accrediting agencies in order to continue to provide healthcare services.

The list goes on.

So, a content strategist in the healthcare industry has to make a lot of decisions about what the right content is for a particular audience. And the main determining factor in these decisions has to be an organization's goals. If a hospital is trying to improve the experiences their patients have, then the right content might be medical information delivered in plain language that ordinary people can understand. If a small doctor's office is trying to make people aware of the specialty services they provide, then that content may be digital content posted on the organization's website. If an insurer is trying to assess the likelihood that certain types of patients will use a new incentive program to lower their risk of heart disease in exchange for cheaper health insurance, then the right content might be an email newsletter delivered to every person they cover that details the benefits of the program.

Often, the right content is a combination of more than one type of content. This is because organizations typically have more than one audience they're trying to reach and more than one channel of communication they need to use to reach them.

For the Right People

Next, content strategists need to determine who the right people are. In this sense, *the right people are the audiences that a specific organization is trying to provide content to for a particular purpose.* Using our healthcare example, there are lots of different audiences that the organizations who make up the healthcare industry may want to target.

These audiences include:

- Healthcare providers (doctors, physician's assistants, nurses, nursing assistants, and technical specialists such as radiology techs)
- Healthcare insurers (insurance brokers, insurance carriers, risk assessment professionals, policy underwriters, and staff who run insurance organizations)
- Patients (people who are receiving actual treatment)
- People in need of medical care (people who have yet to receive treatment for some reason, such as financial cost)
- Families of patients
- Families with sick members who are unable to obtain medical care

Say you are a company whose goal is to reach people who are unable to afford health insurance. There are many companies that have arisen in the US due to the spiraling costs of healthcare that try to provide insurance coverage to people who can't afford it. An example of such a company is Sidecar Health: *https://sidecarhealthinsurance.com/*. Their goal is to provide a better customer experience for people who want additional flexibility and affordability in their coverage (*https://sidecarhealthinsurance.com/how-it-works*).

This content is particular to the US, of course, where we have an insurance industry that is largely privatized. In Europe, Australia, or other countries with socialized or non-profit healthcare, this content may fall on deaf ears.

Content always needs to be *localized* to the specific language and culture of its intended audience, you see. Localization is thus the process of taking content from one culture and translating it for another culture, including not only changing the language from something like English to Spanish, but also translating the content for the different *values* that members of another culture hold.

This sometimes needs to happen within cultures, too. The US is a country in which many different languages are spoken, for instance. It's becoming increasingly important to translate all public notices into Spanish, which is spoken in about 13% of households (Thompson, 2021).

To learn more about localization, see Chapter 12 where we discuss this topic in-depth.

It is safe to say, then, that a primary audience for this company is people who are not covered by a group insurance plan through their employer. Such people, which include professionals like college professors, K-12 educators, doctors, software engineers, and lawyers, typically receive healthcare coverage at an affordable rate through their employing organization. Such organizations are able to insure hundreds or thousands of employees at a greatly reduced rate by purchasing a group plan from an insurance company.

However, at the time we're writing this, there are over 30 million Americans without any form of health insurance (Finegold, et al, 2021). It's a safe bet that a company like Sidecar Health would want to target people from this group with their content much more than they would want to target an audience like lawyers, who probably have affordable health insurance coverage. Uninsured Americans are the right people for Sidecar Health, in other words, because they are an audience that matches up with their organization's goals: to provide affordable healthcare coverage to people who need it.

At the Right Time

In the age of social media, interactive websites, mobile apps, and the ever-expanding smart device market, timing is also a key consideration for content strategists. Content production cycles are becoming faster and faster. Some content, such as social media content, needs to be produced and delivered very quickly, sometimes within seconds of a triggering event. Other types of content, such as healthcare advocacy standards, may take weeks, months, or even years to develop and deliver. And there are literally hundreds (if not thousands) of different types of content that exist in-between, some more temporary, some longer lasting.

As with everything else a content strategist deals with, the issue of timing is largely a factor of organizational goals, audience goals, and content type. The timing of the same type of content, such as a social media campaign centered around patient advocacy, is going to be very different if the goal is to reach patients and their families who have limited access to the internet and thus only check it once a day in the evenings at a local public library.

See what we did there? The more you know about an audience—their reading habits, their levels of education, what technologies they use, what expertise they have, what languages they speak—the more you can tailor content to them. And this brings up a very important topic that we will deal with throughout this book: data.

One of the major advantages of current communication technologies available to content strategists is that many of them automatically collect data on the people that use them. If you install Google Analytics on your organization's website, you can track how many people visit your website, where they are accessing it from, what pages they visit, and how long they stay on each page (Figure 2.1).

That being said, many content strategists also choose to collect more data, such as completing *content inventories* that attempt to capture all the content

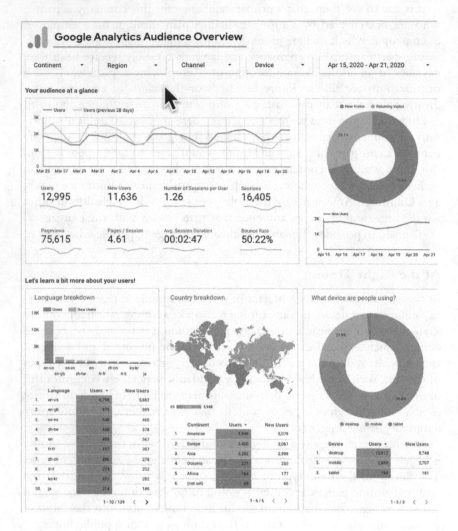

Figure 2.1 Example Google Analytics dashboard

Source: Google and Google Analytics are trademarks of Google LLC and this book is not endorsed by or affiliated with Google in any way.

https://support.google.com/analytics/answer/9849873?hl=en

within a specific organization or channel. This form of research is usually done in service of a *content audit,* or *comprehensive assessment of a collection of content based on predefined criteria.* Many also collect more qualitative data directly from audience members, such as data from interviews and focus groups about specific needs and pain points.

Regardless of the way it's collected, data is very important for a content strategist's job, because it gives him or her important information about how to reach an audience, the needs that audience has, and how often that audience needs new information. This helps a content strategist develop a plan to deliver content in a timely manner.

For the Right Reasons

When deciding on the reasons for developing content for a particular audience, however, things get murkier. The right reasons for developing content can vary widely from organization to organization. One purpose of a company like Sidecar Health is to provide services to customers in exchange for a fee. If they can't accomplish that goal, then they will cease to exist as an organization.

A content strategist working for such a company needs to ensure that their content aligns with that goal. Maybe they don't need to create every piece of content with that goal in mind. There are probably other goals that Sidecar has that we're not aware of as people looking at the organization from the outside. But if their main goal is to acquire customers, then a content strategist who works with them needs to ensure that all the content they create "pushes the needle" of that goal.

At the same time, content strategists that fail to keep the goals and values of their audiences in mind won't deliver successful content. Content goals exist in that sweet spot between organizational goals and audience goals, in other words, as depicted here (Figure 2.2).

Content Strategy Goals

Organizatonal Goals   Content Goals   Audience Goals

Figure 2.2 Content strategy goals

A different organization in the healthcare space, such as a non-profit like the National Health Council, will have very different content goals than Sidecar Health. The primary goal of the NHC is patient advocacy. This organization wants to "provide a united voice for the 160 million people living with chronic diseases and disabilities and their family caregivers" (National Health Council, 2021).

Their reasons for creating content include:

- Influencing policymakers
- Educating patients and caregivers
- Bringing other patient-centered organizations together
- Promoting increased access to affordable healthcare

It's not that Sidecar Health is wrong and the NHC is right, or vice versa. It's more accurate to say that these two organizations have different definitions of "right." Their reasons for creating and delivering content to their target audiences are very different, because their organizational goals and audience goals are very different. And thus, their content goals are very different.

Individual Content Strategy Processes

So, that's a broad overview of the full content strategy process and its ultimate purpose, but how do content strategists go about actually developing and delivering the right content to the right people at the right time and for the right reasons?

Just like any writing or communication process, there are a series of steps they go through, which we turn to now.

Audience Analysis

A very important process for content strategy is audience analysis, or *the process of assessing an audience's needs and pain points* (or moments of frustration during a process). The purpose of this process is to determine, in as much detail as possible, what makes a particular type of audience tick. What are they looking for in the content? When do they want it? What do they value about the content? What problems are they trying to solve by using the content?

This analysis is typically conducted through a combination of the data gathering techniques we mentioned earlier: analytics, interviews, and focus groups.

One of the primary deliverables from this process is a persona, or archetypal audience member. We mentioned personas in the introduction. They're essentially profiles of a particular segment of an audience and include information such as the following:

- Name
- Photo

- Demographics (age, race, gender, location, occupation)
- Story: What makes them a good audience member for this content? What cultural values do they bring to the content?
- Goals and challenges: What is the audience member trying to accomplish with the help of this content? What pain points are they experiencing that can be alleviated through useful information?
- How I can help: What can the strategist do through their content to help the audience member achieve their goals and alleviate their pain?

See an example persona in Chapter 3.

Identifying Content Types and Channels

Another process content strategists engage in is the identification of specific content types that align with specific channels. Once you know the type of information an audience wants, you need to figure out what channel and what content type will be used to deliver that information to them. This is kind of a chicken or egg process, meaning sometimes the channel is determined first and sometimes the content type is.

If your persona is new employees who work for an educational non-profit that delivers after-school programs (our example from the introduction), for instance, then chances are your channel will be internal to the organization. And if this audience primarily needs to learn how to actually run after-school programs, then you're probably going to deliver instructional content that exists in a single place they can access, such as a password-protected file sharing system, internal server, or other database.

The content type may be a PDF they can download and read, if you think the content will rarely change. Or it might exist in a structured form so that it can be easily edited as new programs are added. If the content type is the latter, structured instructional content that can be accessed by a variety of users at once, then it will probably exist in some kind of internal server or other database and have several different types of permission. One type of permission will allow our primary user, the new employee, to access and read the content. Another type of permission will allow developers of the content, such as educational specialists who design the after-school programs and write them up, to make edits to the content. And a final type of permission will allow a content strategist to manage all the instructional content within the organization to ensure that it is up to date.

This is just one example of how a content strategist could drill down to a specific content type and channel from a general audience need. Typically, within a single organization, content is repurposed and shared by a lot of different audiences.

Content Auditing

As mentioned earlier, when we talked about figuring out the right timing for content, content auditing is a tried-and-true research method that content strategists use to assess content. Typically, a content audit begins by inventorying all the content within an organization, or sometimes all the content of a certain type (e.g., website content). This is typically done with the use of a spreadsheet in which the content strategist tracks information like the following:

- Location (e.g., link to a webpage, physical location)
- First created (e.g., the original date when this content was created)
- Last updated (e.g., the last time this content was changed)
- Content type (e.g., blog, structured article, PDF user manual)

From this initial inventory, content strategists assess content based on prede-fined criteria that align organizational goals and audience goals into content goals. These criteria may include:

- Currency: How current is the content? Is it outdated or up to date?
- Authoritativeness: Is the content still authoritative? Does it include information that is still respected?
- Accuracy: Is the content accurate? Has it been recently fact-checked? Does it represent information that is correct and verifiable?
- Audience appropriateness: Is the content appropriate, given the audience(s) it targets? Does it adequately meet their needs? Does it address their pain points and values?
- Visual appeal or design: Is the content effective from a visual stand-point? Does it match current best practices for the design of this type of content?
- Interaction design: How can an audience interact with the content? What actions can they take based on the content? And do these actions seem appropriate to this content type and the type of audience it is targeting?

There are many kinds of criteria that can be used during a content audit. The most important thing is that the criteria match the content goals, or the overall purpose of the audit. We explore the development of assessment criteria for a content audit more fully in Chapter 5.

Assembling a Content Strategy Plan

Content strategists also assemble content strategy plans, or roadmaps for explaining the goals, audiences, channels, and other elements of a specific strategy. These documents are typically used by content strategists to explain the strategy part of their jobs to others within their organization.

Content strategists are like managers of people, audiences, and content. They need to make sure that content creators within the organization are hewing to specific content goals. They need to make sure audience needs are being served. They need to make sure goals are being met, and when they are met, that new goals are set. They need to make sure that content is continually assessed and updated as needed.

To communicate this to other members of their organization, they create detailed plans, typically by drafting a single document that explains all the ins and outs of content. There are many different ways to develop a content strategy plan, but at its core, such a document contains:

- Goals: measurable, achievable, simple, task-oriented objectives for what content should do for an audience
- Audiences: a descriptive rendering of each audience group as a persona or representative audience member
- Channels: a complete listing of all the various communication channels through which content will be delivered
- Content models: frameworks for messaging within each channel that align the specific requirements of the channel with the goals and audiences being targeted
- Editorial calendars: lists of tasks that need to be accomplished on a regular basis, including which tools are best suited for which tasks

---

Content strategy plans can take a lot of different forms and are typically developed to suit the needs of a specific organization. However, we explore an example workflow for developing a content strategy plan in Chapter 7.

Collaborating with Other Content Developers

It's also important to note that content strategists collaborate a lot with content developers. Sometimes they develop content on their own, but they also often work with people within an organization who develop content on a regular basis to help these content developers move in the right direction.

This can include:

- Organizing planning meetings
- Training content developers in best practices
- Developing editorial calendars and making sure content developers follow them
- Editing content

- Delivering content created by others
- Working with subject matter experts to glean specialist information from them
- Tracking content goals and adjusting what the organization is doing with content, accordingly

Content strategists are people people as much as they are content people, in other words! They work with lots of different people to ensure that an organization's content performs well.

> Content strategists collaborate with other content developers in lots of different ways, but we explore some of the strategies they use to do so in Chapter 7.

Content Modeling

Another important process content strategists engage is called content modeling, which is essentially *the process of creating structures and frameworks that content must adhere to*. You are probably familiar with a very common form of content model: the outline for an essay. An outline is a content model for the genre of college essay.

This means an outline does the following things:

- Explains the overall structure of the genre
- Lays out a plan for what information should be included in the genre
- Defines genre-specific attributes, such as an introduction containing a thesis statement, body paragraphs that include claims and evidence, and a conclusion that sums up the argument and provides final thoughts

Content strategists extend this logic by modeling content for the many different genres of communication they help create, be they blog posts, structured articles in a database, or questions and answers posted to a user forum. The advantage of content modeling is replicability: the content strategist defines the purpose, goals, and components of a specific content genre (or *a collection of individual content types*) to ensure content creators produce future versions of that genre to spec.

Content strategists are busy people, you see. They live and breathe content day-in and day-out for the organizations that employ them. They may manage or create hundreds to thousands of pages of content a month. And imagine if they created all that content only to discover that it wasn't structured properly!

You might have experienced that if you have ever skipped the outline for an essay for one of your classes and jumped right into writing your essay. Maybe you're the type of writer who can do that effectively, maybe not.

Chances are, you forgot one or more elements (such as the evidence for one of your claims) and had to go back to add these elements. Maybe you didn't even notice you had failed to include these elements until you received your final grade. The advantage of creating an outline is to ensure you know all the information your genre (a college essay) should contain *before* you start working on it in earnest.

Now, imagine if your job was to create fifty or one hundred or two hundred college essays per month. And imagine if you didn't have a clear outline of what each of those essays should contain. You might be adding hundreds of hours of extra work into your schedule that could've been saved through the use of a complete outline for what each essay should contain.

This is the power of content modeling: planning ahead so that your content is to spec when you deliver it.

> Like all the parts of the content strategy process, content models can take many forms. However, we provide some common content models, as well as a simple plan for creating content models, in Chapter 6.

Revising and Editing Genres

Revising and editing are an important part of content strategy, too. No one is perfect. Even the best-made content model might contain flaws, for example, or a given content developer might fail to follow some aspect of it. This is why regular review of all content is essential *before* it gets published.

Often there are several levels of review that happen to content within a successful organization:

- Formative review: reviews of initial drafts of genres to make sure they are headed in the right direction
- Summative review: reviews of semi-finished drafts of genres to make sure they meet basic quality standards
- Editorial review: corrective reviews that ensure content perfectly matches organizational guidelines for tone, style, audience appropriateness, and grammar

Even with effective content models in place, creating a wide variety of communication deliverables is taxing work. Content developers sometimes get distracted, forget things, or simply fail to understand what's expected of them. If no one is checking their content before it gets delivered to its target audience, chances are the content will be sub-par.

Content, you see, is an important product within an organization. Just like any other product, be it an axle for a car or the next installment in a bestselling novel series, it needs to be reviewed for quality before being passed on to the consumer. And typically it needs to be reviewed multiple times to *ensure* the consumer doesn't receive a faulty product, or, in this case, ineffective content.

> We discuss tactics for revising and editing different genres of content in Chapter 9.

Ensuring Content Usability and Accessibility

Because there are so many different content types and genres that today's organizations use, from user profiles displayed in a mobile app to an interactive help forum that uses chatbots to guide a customer to the answer to their question, content strategists also need to ensure that their content is usable and accessible.

Let's break down these two words, as they mean very different things. According to Jakob Nielsen, a leading user experience expert, usability is a "quality attribute that assesses how easy user interfaces are to use" (Nielsen, 2012). When we talk about *content usability*, we're talking about *the usability of a specific piece of content*, not the interface of the device it's displayed in, per se.

So, content strategists might assess how easy it is for audience members to find answers to their questions in a help forum, or if they can learn how to use a new piece of technology based on an article published to the manufacturer's website.

Accessibility, on the other hand, refers to *the ability of people with some kind of impairment, often from a disability, to successfully use information*. Say you are a content strategist working for a non-profit that delivers important patient information to people with visual impairments, for example. If your content is not developed correctly so that it can be read by a screen reader, then it will not be accessible to this audience, who might not be able to successfully read it, even with the use of assistive technology.

> We discuss some of the techniques for ensuring content usability and accessibility in Chapter 10.

Delivering, Governing, and Maintaining Genres

Many genres of communication now exist in a form that needs to be updated. This is particularly true with digital content but can also be the case with paper content as well. If a webpage meant to inform customers about new versions of a specific product isn't updated, then it will fail in its intended purpose. And it's easy for digital content, in particular, to quickly become outdated.

This is why a primary job of many content strategists is ensuring that content doesn't become outdated.

This entails three primary forms of activity:

- Delivery: Content strategists need to ensure that their content actually reaches its intended audience in the correct form.
- Governance: They need to also ensure that this content is managed effectively so that it keeps doing what it needs to do or is retired if it's outlived its usefulness.
- Maintenance: They need to maintain this content by checking, sometimes in real time, that the content meets basic quality standards.

The most difficult thing about the delivery, governance, and maintenance of content is setting up a system ahead of time for managing content throughout its entire lifecycle. This should ideally start at the planning stages and should continue as long as the content is accessible to even a single audience member.

Continual content audits can be an invaluable tool in this process as they can help content strategists keep on top of necessary changes to content and can also indicate when it's time to remove content.

We discuss specific strategies for delivering, governing, and maintaining genres in Chapter 11.

Localizing Content

Yet another important activity is the process of *localizing content* or *ensuring that content is effective within a specific cultural context.* This is important because the content of many organizations is now global in scope: audiences from all over the world can access it, use it, and reject it if it doesn't perform to their expectations.

The main challenge in this process is understanding the idioms of specific cultures and the conventions these idioms entail. Certain colors, for example, have very different meanings across cultures. Certain terms that are perfectly understandable to readers of American English might be completely confusing to readers of British or Australian English. Even within different regions of the same country, there can be important local inflections to take into account.

Content strategists who work to localize content often work with translators and cultural insiders to adapt content to multiple contexts. Think about a content strategist working for a large, multinational corporation like IBM. IBM delivers many different types of content to people from hundreds of different cultures. What works in Berlin doesn't work in Belize City. Rather, content strategists within IBM need to work closely with specialists in both Berlin and Belize City to ensure that meaning, usability, accessibility, and overall communication remain intact.

Localization is a broad topic that has entire books of its own. However, we discuss some of the common strategies for localizing content in Chapter 12.

Using Tools and Technologies

Last, but not least, content strategists use a variety of technologies in their trade. From web crawlers that can quickly tell them how many pages are within a specific website, and when they were last published, to content management systems that allow them to easily manage, revise, edit, and publish content in a variety of genres, these technologies enable content feats that were unheard of even a few years ago. These feats include the simultaneous delivery of content to thousands of users all over the world, version control that tracks every change made to content over its entire lifecycle, and the capturing of various forms of user data that provide an accurate picture of how content is being accessed, used, and repurposed.

It's difficult to talk about technologies adequately in an introductory book because they change so quickly, but we think it's important to do so because they are such a big part of what nearly every content strategist does.

As we mentioned in the introduction, these tools can take a variety of forms, depending on the specific duties of individual content strategists. These forms include:

- Desktop publishing software
- Collaboration tools such as video conferencing software and collaborative writing technologies
- Authoring tools that allow writers to quickly create content in a basic format and then output that content into a variety of other formats
- Open-source information architecture that helps writers structure content in such a way that it can be used by a variety of other people and technologies
- Content management systems (CMSs) or technologies that automatically format and store content for future use
- Component-based content management systems (CCMSs) that break down content into its basic components so that it can be reassembled later into complex genres such as large-scale manuals that need continuous updates
- Application programming interfaces (APIs) or tools within existing applications that enable writers to build their own simpler applications, often for the purposes of storing content within those larger applications

Though an exhaustive list of all the technologies content strategists use is beyond the scope of a single book chapter, we discuss some of the common types of technologies, including examples, in Chapter 13.

Heuristics, Not Linear Steps

After this brief overview of the content strategy processes, you may be feeling overwhelmed. You might be saying to yourself: "no one professional can do all that!" And you would be right in thinking that content strategists are jacks and jills of many trades. It is also true, however, that individual content strategists engage in different duties across organizations. There is little, if any, standardization across this field.

This means that there are no rigidly and universally-accepted best practices for many of the processes we'll describe in this book. They are based on our extensive research, and practice, in this field, as well as conversations with many content strategists and content strategy teachers.

Much less a series of linear steps, try to think of these processes as heuristics, or *means of discovering or learning things for yourself.* In your journey to learn about content strategy, you may discover that you're really good at content auditing, for example, and want to focus on that. You may land a job that is very data-intensive and requires you to get really good at working with analytics. Or you may work on a highly collaborative team of content developers that each specializes in a different part of the overarching content strategy process.

One thing that is certain about content strategy is that it's a wide-open field with a lot of opportunities for innovation. And many of the content strategists we've personally spoken to over the years started out in a very different role, such as a technical writer, but saw a need within their organization and pursued it. Content strategists are nothing if not self-starters, in other words. They take on difficult problems that others dismiss as too complex and find solutions that help organizations meet and exceed their content goals.

References

About. National Health Council. (2021, June 11). Retrieved September 16, 2021, from https://nationalhealthcouncil.org/about/

Finegold, K., Conmy, A., Chu, R. C., Bosworth, A., & Sommers, B. D. (2021, February). *Trends in the U.S. uninsured population, 2010–2020.* ASPE. Retrieved September 16, 2021, from https://aspe.hhs.gov/sites/default/files/private/pdf/265041/trends-in-the-us-uninsured.pdf

Nielsen, J. (2012, January 3). *Usability 101: Introduction to usability.* Nielsen Norman Group. Retrieved September 16, 2021, from www.nngroup.com/articles/usability-101-introduction-to-usability/

Thompson, S. (2021, May 27). *The U.S. has the second-largest population of spanish speakers-How to equip your brand to serve them.* Forbes. Retrieved September 16, 2021, from www.forbes.com/sites/soniathompson/2021/05/27/the-us-has-the-second-largest-population-of-spanish-speakers-how-to-equip-your-brand-to-serve-them/

Further Reading

Casey, M. (2015). *The content strategy toolkit: Methods, guidelines, and templates for getting content right*. New Riders.

Department of Health and Human Services. (2016, January 24). *Content strategy basics*. Usability.gov. Retrieved September 16, 2021, from www.usability.gov/what-and-why/content-strategy.html

Halvorson, K. & Rach, M. (2012). *Content strategy for the web*. 2nd ed. New Riders.

Larry, says, D. T. & Thackeray, D. (2019, March 5). *A content strategy process model: How content strategy works*. Elless Media. Retrieved September 16, 2021, from https://elless-media.com/content-strategy/process-model/

Michels, L. (2019, February 8). *The five stages of successful content strategy*. Passion Digital. Retrieved September 16, 2021, from https://passion.digital/blog/the-five-stages-of-successful-content-strategy/

Getting Started Guide: Exploring the Content Strategy Process

The content strategy process we've presented in this chapter is our synthesis of a lot of different sources and our own experiences doing content strategy. Many content strategists disagree about what the content strategy process entails.

Try doing some of your own research into the different ways content strategists describe this process. Review one or more of the sources listed in the Further Reading section. Also, try using Google (or your other favorite search engine) to search for phrases like:

- What does a content strategy look like?
- Content strategy process
- Content strategy development process

Peruse the information you find and use it to answer some of the following questions about the topics presented in this chapter:

1. How is the content strategy process described in the sources you found? How do these descriptions compare to the overall process we've described in this chapter?
2. What elements of the content strategy process did you come across that weren't mentioned in this chapter? What is the purpose of these additional elements, according to the authors you located?
3. Did you find any information that is specific to the content strategy process within a particular industry or context (e.g., healthcare, the sciences, education, engineering)? How does this information compare with our generalized discussion of the content strategy process in this chapter?

3 Audience Analysis

Now that we've gotten a basic understanding as to what a content strategy workflow looks like, we can start to work on some of the research that will allow us to create the most effective content strategy. As mentioned in the previous chapter, this means *thinking about who the right people are that we are trying to deliver content to for a particular purpose.*

But how do we know who the *right people* are?

Our first task in answering this question is identifying the actual, real, live people who are going to be reading, interacting, and engaging with our content for a particular purpose. Depending on who you talk to, different kinds of professionals may refer to these people as users, consumers, stakeholders, or a whole host of other terms. However, for content strategists, the most common term is definitely *audience*. An audience can be defined as the *groups of people who an organization is trying to reach with its content.*

Notice we said groups, not group. An organization's audience is always composed of different groups of people. If modern-day organizations were only trying to reach a single, specific group of people, then their content goals would be much simpler and content strategists might not even be needed. But the people who interact with today's organizations almost always have diverse needs, goals, and pain points. They may access content on different devices, in different languages, and using different levels of knowledge. They represent different genders, different sexual preferences, different races, and different socioeconomic classes.

This is why content strategy almost always begins with audience analysis, with researching the details of an organization's audience to determine the demographics, goals, and pain points of the people who interact with the organization's content.

What Is Audience Analysis and Why Is It Important?

Audience analysis is a common methodology utilized by many professionals including technical communicators, user researchers, and content strategists in order to identify key attributes of the people they are developing content for. Gone are the days when organizations could afford to blast out generic information and reach a wide audience. There's simply too much content out there to distract modern consumers. The secret to reaching a wide audience in the days of mobile devices, seemingly omnipresent internet connectivity, and an ever-increasing array of content types, is to be *specific*, not generic. The more you know about your audience, the better.

This is why content strategists typically do audience analysis before they do anything else. When you think about it, this makes a lot of sense. How can we focus on creating and delivering the best content to a specific group of people if we don't know who those people are? We need to know things like who they are, what they want from our content, where they like to get content from, and how we can deliver content to them that will really capture their attention.

At the same time, in our experience as sometime content strategy practitioners, your typical organization does *not* have a firm grasp on its audience. Out of all the clients we've worked with over the years, we have yet to encounter one who had a specific, identifiable audience in mind for their content. When we ask who they're trying to reach, clients typically mention the role the person plays, such as "customers," "donors," "volunteers," or "employees." Sometimes they may know a bit of demographic information, such as "men between the ages of 35 and 65," but that's about as specific as it gets.

And many of our clients have been actively resistant to narrowing down their audience to specific, identifiable groups. The perception by people in many organizations is that if you narrow your audience, you'll reach fewer people. That simply isn't how modern audiences consume content, however.

Think about your own content habits for a second. When you are looking for something, such as a particular product or service, where do you go? Do you just browse idly online, or do you immediately go to a specific source of information? And when you do consume information about a product or service, do you trust all forms of information equally? Do you trust marketing copy on a company's website as much as you trust reviews by customers?

If you're like the vast majority of modern consumers, your answers to these questions show that you are pretty picky when it comes to content. In fact, research by companies that study consumer habits, such as McKinsey & Company, a world-wide leader in this area, shows that consumers are very fickle when it comes to buying (*Ten Years*, n.d.). We have access to so much more content before making a purchase today than we did even ten years ago, that we've become expert shoppers. The bigger and more expensive the purchase, the more information we gather.

And our loyalty as shoppers is much harder for organizations to gain. If we find a better deal, better reviews, better *information* elsewhere, we're likely to abandon even a trusted organization for another provider. In fact, 58% of consumers change brands from one purchase cycle to the next (*Ten Years*, n.d.).

And a big factor as to whether or not we stay loyal to a particular organization, besides what content we consume during our initial research, is post-purchase support (Court, et al, 2009). Not only do we want to have a ton of information about a product or service before we make a purchase, but we want to be catered to after our purchase as well. We want to be taken care of during our entire consumer journey, in other words, and we want to be actively courted.

Given all this, perhaps you understand a little bit better now why an organization needs to know its audience inside and out. It's not enough to provide a good product or service anymore. You have to court consumers and demonstrate, over and over again, why you're the right provider for them.

Identifying Potential Audiences

Now that you understand a bit about what's at stake when it comes to modern audiences, it's also important to understand that not all audience members are created equal. Rather, there are many different types of audience members. And one key difference in audience members has to do with whether they are internal or external to an organization. When we think about delivering content to an internal audience, we're thinking about creating content for people that work within an organization, such as employees or volunteers. When we think about reaching an external audience, we're thinking about people who are outside of the organization, such as clients, donors, or customers.

And while any audience we're trying to reach, be it internal or external, must be thoroughly researched, often using similar methods, internal and external audiences typically have very different needs.

Types of Internal Audience Members

Audiences that are internal to organizations are often just as diverse as their external counterparts. Whether we are developing content to deliver to coworkers, superiors, or investors, each audience member is a unique individual with specific needs.

Although there are many different roles, positions, and job titles in modern organizations, some common types of audience members include:

- Executives
- Managers
- Marketers
- Salespeople
- Content specialists (writers, editors, reviewers, etc.)
- Support specialists (customer support, technology support, subject matter experts, etc.)
- Design specialists (mobile, web, graphic, video, etc.)
- Development specialists (product developers, engineers, programmers, etc.)

WHAT INTERNAL AUDIENCES NEED FROM THEIR CONTENT?

Depending on the type of internal audience member you're trying to reach, content needs of individuals such as these can vary widely. Say you're a content strategist for a company that is getting ready to deliver a year-end newsletter to all of their current subscribers. While you and your team are focused on developing meaningful content for the newsletter, you are also collaborating with members from another team—the design team—to make sure that the newsletter looks visually appealing and matches the company's brand.

In this scenario, members of the design team have very specific needs. They probably need to know all the different content types so they can design a layout for the newsletter. They need to know what color scheme you're picturing for the newsletter and how this scheme will clearly extend from the company's overall brand guidelines. They may need details of specific calls to action, including how text, images, and graphics should align to compel newsletter recipients to take actions such as making a purchase, engaging with other types of content, or reaching out for support regarding a past purchase.

While this is only one example of how you might interact with an internal audience as a content strategist, hopefully it gives you some context for how you might do so. We talk about internal audience members throughout this book, including how to collaborate with them.

TYPES OF EXTERNAL AUDIENCE MEMBERS

Much like internal audiences, external audience members occupy many different roles when they interact with content. Some common roles include:

- Clients
- Customers
- Donors
- Potential volunteers
- Potential employees
- Potential investors

It should also be noted that while many of these external audiences will stay external to an organization, there are some that begin as external audience members but may become internal, such as potential volunteers, employees, and investors. There's definitely more of a dotted line between internal and external audiences, rather than a solid line, in other words.

WHAT EXTERNAL AUDIENCES NEED FROM THEIR CONTENT?

External audience member needs will also vary, but what sets them apart from internal audiences is lack of familiarity with an organization. Even people who have engaged with an organization on an ongoing basis, such as returning customers, aren't inside the organization to see its innerworkings. Again, however, some external audiences can become internal over time, such as a potential volunteer who hears about a non-profit and decides to get involved.

Regardless: anyone who is not actively working within an organization at any given time can safely be considered an external member. This means that you have to treat them as though they don't know the central processes, services, products, and specialized terms of your organization. You don't want to insult the intelligence of your external audience members, but assuming they can find information on your company's website when they need it just like an internal audience member would be able to can be a fatal mistake.

Part of content strategy consulting, at least the way we do it, is sitting down with someone from an organization to discuss how they can improve their content. During one of these initial meetings, we asked a marketing manager for a small, regional non-profit who the primary audience for their website was. As the purpose of their organization was to improve the downtown region of the city they were located in, the manager confidently said: "Investors! We want to attract investors who will repurpose property and rent spaces downtown."

We were surprised by this, because the organization's website, which we had reviewed in preparation for the meeting, had no language about investors that we had located through a cursory navigation. We relayed this information to the manager and her brow furrowed.

"Oh, that's not right. Let me show you!" She said.

So, one of us turned our laptop to face her and she instructed us how to get to the investor's portion of the website. It went something like: "Ok, click there. No, not there. There! Yes, that's right. Okay, and then click there. Then there's a dropdown. And . . . yep! There it is. Right there! It's obvious!"

This was several years ago and we didn't end up working with the non-profit, but I'll never forget this encounter because of how shocking it was that a member of an organization would think that it was "obvious" to click several pages into a website to find information for its primary audience. Don't get us wrong: there was nothing wrong with this marketing manager. She was a bit on the young side, if I recall, and this was her first job after college, but I had actually met her several times and she had actually impressed me with how knowledgeable she was about content and how it works.

It's just *that easy* to forget that external audiences have no idea where to look for your content that you think is obviously placed.

This is a hallmark of the field of user experience (UX), a principle that many refer to as: "[y]ou are not the user" (Babich, 2020). The principle says that as people who design things, it's easy for a UX designer to assume that their user understands the product they're designing in the same way that they themselves do. But this is a huge mistake. In fact, the user of a product *doesn't* understand the product the same way the designer of a product does. Not at all.

We could also apply this principle to content strategy, by saying *you are not your external audience.* As a content developer, you may think your content is *obviously* and strategically placed. You may think that it's the best content ever created: clear, concise, engaging, easy to understand. But the question you should be asking yourself is: *what does your audience think?*

So, what we're really saying here is that it's always important to think about what your audience needs, regardless of whether they are internal or external. As a content strategist, you need to learn about your audience and tailor content accordingly. Often that means doing some research in order to learn more.

Methods for Gathering Audience Data   While there are many different research methods that we could deploy to gather information about our audience, we want to focus on the methods we have found to be most useful for doing audience analysis (and the methods that many other content strategists have found to be useful as well). This section will thus focus on analytics, interviews, and focus groups.

Analytics   The word *analytics* may be scary if you're new to gathering audience data online. Using analytics in order to identify our audience isn't as complicated as it may sound, however. In fact, many of the platforms that you are probably using to publish your content (such as many social media platforms and website content management systems like WordPress) already have analytics features at your disposal. What makes this a useful method is that you don't have to go and find your audience out in the wild. You don't have to track them down. You don't have to set up times to meet with them. You can just look at the passive insights that have been collected already for you and start to utilize that information to craft better content.

In fact, if your organization has already published content to social media channels (see, for example, Facebook's IQ tool: https://www.facebook.com/business/insights/tools/audience-insights) or a website (e.g., Google Analytics: https://analytics.google.com/analytics/web/), you will be able to go into your organization's accounts and look at important information about your audience, including the demographic information about all the people that currently follow you or visit your website. Even if your website doesn't have analytics set up on it currently, this can be done quite easily and the data can still be gathered.

Analytics essentially allow a content strategist to see data on how people are interacting with your content. Unfortunately, there's no universal standard for how this data is collected, analyzed, or represented to users of an analytics dashboard. This means that if you view analytics on Facebook using the native Facebook analytics tool, IQ, then view this same data using a third-party tool like Hootsuite (https://www.hootsuite.com/platform/analyze), there's no guarantee that the data is being collected, analyzed, and represented in exactly the same way.

But that's okay. Content strategists aren't data scientists. That's a whole other job. The point is, in many cases, depending on the platform, you can tell several things from analytics, such as:

- If people are engaging with your content and instantly abandoning it (bounce rate)
- What content people engage with the most
- What content people tend to ignore
- Some of the demographics of users, including their location, their gender, their occupation, and what keywords they used to find your content

Analytics aren't answers in and of themselves, in other words, but they enable content strategists to ask important questions that could help us create more impactful content moving forward. For example:

- What are some of the demographics of audience members?
- Are any of these demographics important to your understanding of your audience?
- Or better yet, is any of this information new or surprising to you?
- Does this information help you craft better content moving forward?
- What specific genres of content are most popular? Blog posts? Videos? Polls?
- What kind of content got the most engagement?

Maybe you and your team believed that visitors to your website were mostly interested in content related to investors, but actually they're interested in making a donation. This could be because your investor-related content isn't strategically placed. Or it could be because investors simply don't know you exist.

Often, when we are looking at a client's analytics, they generate more questions than answers, but that's okay. That's why we have other research methods we can employ to try to answer those questions.

If you want to see an example of what an analytics dashboard looks like, see Figure 2.1 in Chapter 2 for a screenshot of the Google Analytics dashboard.

If you're looking to get started with analytics right away, you can skip ahead to Chapter 13, the chapter on tools and technologies, where we present several popular options for gathering analytics on websites, social media, and several other channels.

Interviews   Interviews are one of the most common research methods for gathering data on audiences. They allow you to do a deep dive with individuals who fit the demographics of the people you're trying to reach. They help elicit stories, understand goals, and define pain points that people are experiencing in regard to your content.

Conducting interviews with internal and external audiences are very different, however.

INTERNAL

The goal of interviewing internal audiences is often process improvement. Internal audiences are on the same team as you, in that individual audience members are also working toward the collective goals of your organization. So, the goal of your content for an audience like this is often to make this work easier and more efficient.

In the example we used earlier of our small non-profit, the marketing manager could have realized the organization's website was not well organized if the goal was to attract investors. She could have approached the Executive Director of the organization with this information and discussed a plan to reorganize the website (we say *could have* because we actually peaked at the organization's website while writing this book and their investor content is still buried beneath several subpages, sigh).

If she had done so, then the organization could have worked with us, some content strategy consultants, to reorganize the website. We've certainly done lots of projects like that before. Our first question to clients who want to reorganize the content on their website is: how do you want us to deliver this reorganized content to you? Do you want it in a spreadsheet? Written out in a document? Do you want us to create a diagram showing where the content should be placed, based on our analysis?

Our job as content strategy consultants is to make things easier on the members of organizations we consult with, in other words. A lot of our work over the years has involved optimizing processes, such as documenting how a marketing manager can add new content to a website built in the Joomla! content management system (CMS).

Here are some sample questions to ask internal audience members of organizations:

- First, we'd like to get some basic demographics from you (ask *pertinent* demographic questions, meaning ask about aspects of their identity that align with the audience you are trying to reach).
- What are the goals of your content?
- Who are the main audiences you're trying to reach with your content?
- How do these goals align with your organization's goals?
- Do you ever feel your content goals aren't aligned with those of your organization? How so?

- What processes do you struggle with the most, in regard to your content (e.g., developing content, managing content, publishing content, tracking the impact of content, understanding audiences)?
- How do you use content most often within your organization?
- Where do you feel your content is currently falling short?
- What channels do you currently use to publish your content (i.e., website, email, social media, intranet or listserv, internal content repository)?

EXTERNAL

When interviewing an external audience member, however, you're often trying to learn more about what they want to get out of an organization's content. External audiences aren't involved in the production of content (with user-generated content being a notable exception—see the following callout box), so they mostly use content to serve their own ends, such as learning more about a product or service, troubleshooting an issue they're having, or making a purchase.

In our running example, if the non-profit had taken our advice and engaged us (or someone!) to reorganize their website, a natural first step would be to interview some local investors. It's hard to attract someone to your website if you know nothing about them. One question we brought up with the marketing manager when we met with her was: where do local investors get their information when they're looking to make an investment? The marketing manager simply blinked at us, then changed the subject back to how well their website was performing, overall.

Again, this is not to bash this particular marketing manager. Many marketing managers at many organizations we've dealt with have made the same mistake: there's an assumption that just because your content is getting hits it's doing what it's supposed to do. However, if you're not even interacting with your audience members in a direct way, can you be sure of that? Have you ever asked them what they think your content is supposed to do?

And even more importantly: have you asked people who *aren't currently members of your audience* what they think? This is another trap that many organizations fall into in our experience, the trap of "well, we already have an audience." We can't tell you (because we literally haven't kept count) how many times we've consulted with someone within an organization who told us that they were doing fine, audience-wise, that they got plenty of traffic to their website, had plenty of subscribers to their newsletter, had plenty of customers, and so on.

After a few questions, however, it became apparent to us that all of these clients (and would-be clients) didn't actually know their audiences that well. Some of them knew basic demographics. Some of them certainly *thought* they knew their audience well and could even rattle off a bunch of facts about them, such as "my typical customer is an African-American man in his 40s who is looking for high-end hair care products." However, every time we asked questions about motivations, behavioral triggers, and what really made their audience members tick, we got a reaction similar to the one the non-profit marketing manager had.

As a rule, the more you understand the motivations of individual audience members, the more you can look for patterns in these motivations. Where do they look for information on a given topic? What triggers them to take action? What triggers them to make a purchase? What pain points are they trying to alleviate in their lives by seeking a particular product or service? What goals are they trying to accomplish by seeking out specific content?

If you can't answer questions like these about an audience, then you don't know them as well as you could, and chances are you are missing out on key insights that can help create more impactful content.

Here are some sample questions to ask external audience members of organizations:

- First, we'd like to get some basic demographics from you (ask *pertinent* demographic questions, meaning ask about aspects of their identity that align with the audience you are trying to reach).
- What are your goals when looking for information on this topic? What are you trying to accomplish?
- What problems are you trying to solve when looking for information on this topic? What pain points are you trying to alleviate?
- Where do you typically go to find information on this topic?
- What channels do you currently use to find information on this topic (i.e., websites, email newsletters, social media, customer reviews, print magazines)?
- What kind of information inspires you to take further action (e.g., subscribing to a newsletter, contacting an organization directly, following an organization on social media, making a purchase, making a donation)?
- How appealing do you find this piece of content?
- How well does this content align with your personal goals for this type of content? Is this content helpful, in other words? Does it do what you expect it to do for you? Why or why not?

As far as how to actually conduct an interview, we largely agree with the tips recommended by Babich (2021):

Step 1: Before the interview

1. Set a clear goal
2. Make sure the interview is the right tool for the problem you want to solve (see more on this later in our focus groups section)
3. Decide who to recruit
4. Design your interview questions
5. Create a good environment

Step 2: During the interview

1. Don't conduct the interview alone
2. Put yourself in a positive mood
3. Stick to the semi-structured interview format
4. Build rapport with interviewee
5. Resist the urge to judge or educate your interviewee
6. Ask permission before audio or video recording
7. Start off with the easy questions
8. Prioritize open-ended questions
9. Ask follow-up questions
10. Use the critical incident method
11. Avoid leading questions
12. Clarify interviewee responses in real time
13. Minimize note-taking
14. Don't be afraid of silence
15. Finish with a wrap-up summary

Step 3: After the interview

1. Conduct a retrospective
2. Structure the information
3. Combine interviews with other techniques

You can read his full article to understand the process of conducting an interview better.

User-generated content is an often ill-understood type of content that creates pain points for both organizations and their audiences at the same time that it solves other problems. If you've used the modern web recently, you've definitely encountered user-generated content. It's everywhere these days: in Facebook posts, user support forums, and wikis created by fandoms.

User-generated content was first introduced during the transition to a "Web 2.0," which promised a more democratic, user-focused web. And boy did it deliver. The problem, as many organizations have now realized, is that not everyone in a democratic space believes there should be rules governing them. Companies like those behind social media platforms quickly realized that they would have to screen out offensive content, such as hate speech, threats of violence, and other kinds of illicit content to avoid alienating their users that just wanted to stay in touch with their friends and family or figure out how to reboot their phone.

Anyone who lived through the 2016 and 2020 U.S. presidential elections knows all the problems this content moderation process has caused. Suddenly social media companies were on the forefront of what was acceptable political speech and what was "fake news." Misinformation spread like wildfire. Several high-level celebrities, politicians, and influencers, including a sitting President of the United States, were banned from several platforms.

This is all to say: organizations have not tempered the storm that user-generated content unleashed. At all. And as more and more users generate more and more content online, companies will continue to face strong challenges from a variety of sectors: political, financial, legal, strategic, technological, and so on.

If you want a deep dive into user-generated content and how it works within an organization, check out the case study two of the authors conducted of an organization that built its whole business model on this type of content (Getto & Labriola, 2016).

Focus Groups   Another method for gathering audience insights that content strategists use is a group interview, also known as a focus group. Much like interviews, your approach will differ depending on whether your audience is internal or external. The reason to conduct a focus group instead of an interview is simple: to gather insights about an audience in a social setting.

People are social creatures. We like to talk to each other. We talk about content we're interested in, organizations we support, organizations we've had bad experiences with. There's a reason that 92% of consumers trust personal referrals: if you know and trust someone, and they've had a good or bad experience with an organization, that's going to color the way you see that organization (Georgiev, 2022).

This is why marketers are often trying to get consumers into focus groups. They don't just want to know how an individual reacts. They want to know how that individual is going to be influenced by other consumers.

So, when deciding whether to conduct a focus group or an interview, you need to ask yourself if you want to understand individual audience member behavior, or if you want to understand how individual audience member behavior *is influenced by others*. With an interview you get a much less biased sample of what motivates individual audience members. With a focus group, your data will be biased by the behavior of other people in the focus group, but this can be useful to understand.

If our would-be non-profit client had discovered that investors actually talk to other investors before making an investment decision, then it would be important to conduct some focus groups. It would be important to understand how groups of investors influence the decisions that individual investors make. If the organization had data indicating that investors were lone wolves who didn't trust what other investors had to say, then it would be better to conduct interviews.

Now, this also means that sometimes it's often better to conduct interviews first. That way you're getting a less biased view of individual motivation. If you conduct several interviews and your participants keep saying that what motivates them is the recommendations of others, however, then it's time to organize some focus groups so you can better understand that dynamic.

You can use similar questions to those we recommended previously for conducting focus groups. You're largely trying to understand the same things, just in a different format.

As far as how to actually conduct a focus group, you can't go wrong by following the guidelines of Thompson (2016):

1. Ensure you have clear objectives
2. Recruit the right people for you
3. Pilot your focus group before the real thing
4. Create a happy atmosphere
5. Keep control of the session
6. Avoid leading questions
7. Rope a colleague in to be your assistant moderator
8. Send participants away feeling positive about their experience

You can read his full article for more information.

INTERNAL VERSUS EXTERNAL

The rationale for conducting focus groups with internal versus external audiences are largely the same as those for conducting internal versus external interviews: it depends on the type of content you're trying to develop or optimize. Are you trying to build a newsletter for employees? For first-time customers? For potential volunteers? You need to seek out people who are similar to the types of people you're trying to reach.

Other Sources of Data You also don't want to neglect types of data that already exist within an organization. At the very least, these types of data can help you generate questions you need to answer about your audience. Sometimes they can also answer questions that you won't need to burden people with during interviews or focus groups, which can be taxing for both busy individuals and overburdened content strategists.

These data include:

• Customer survey results
• Support tickets

- Analytics from various channels (as previously discussed)
- Patterns in user-generated content
- Search patterns
- Usability test results (we talk about these in Chapter 10 if you want to conduct your own)
- Requests for new features

A Workflow for Conducting Interviews and Focus Groups   The first step in actually conducting interviews and focus groups with real, live audience members is to segment your audience. And no, this doesn't involve cutting people in half like a cheesy magician. Rather, segmenting your audience means *dividing it up into discernible groups*.

To brainstorm audience segments, start by thinking about what attributes differentiate individual audience members from one another. You can also think of attributes from the following list you think are important to each segment:

- Industry
- Demographics (age, gender, sexual preference, race, ethnicity, nationality, etc.)
- Experience level
- Buying behavior
- Lifestyle
- Customer type
- Income
- Profession
- Technical expertise
- Education
- Location
- Specific channel usage (e.g., web, social, email)

This is where the other sources of data mentioned previously really shine. Any information you can collect before you actually start interviewing people or conducting focus groups will really help out.

Remember, however, conducting interviews and focus groups is also a fact-finding mission, so you may very well find that your initial audience segments are wrong and need to be revised once you start talking to actual people, but that's okay! Audience analysis is about discovery.

A lot of clients ask us during this process: *how many segments should I have?* This is a hard question to answer. Your segments should be linked to the goals you're trying to achieve through your content. In our running example, investors would be a key segment to interview, but there are many different types of investors. There are real estate investors, small business owners, venture capitalists looking to fund startups, and so on. If the main audience the organization is trying to attract is investors, you'd probably want to break that audience up into smaller segments to investigate what type of investor is actually interested in the organization.

Next, you need to brainstorm your interview or focus group questions. See the preceding sections for tips on doing that. You definitely need those questions in front of you as you conduct your research, however.

Once you have initial segments and interview questions brainstormed, it's time to recruit actual people. The only way we've ever been able to recruit interview and focus group participants is guerilla style. For external audiences, this means that we think of people we personally know who fit the target demographics of a potential audience member but *who don't know about the organization we're working with*. This last part is key, because if you contact people who are already familiar with the organization for which you're doing audience analysis, you'll get a biased sample. People who are already familiar with an organization's content already have a lot of opinions about it. You want a fresh, unbiased perspective.

For internal audiences, the recruitment process is much simpler: you need to contact people from within the organization who match the target demographics of the audience you're analyzing. If you're helping the organization develop content for the team responsible for the organization's website, you'll just contact members of *the team responsible for the organization's website*.

Regardless, you want to brainstorm a list of possible contacts. People are busy. Not everyone will agree to participate in your research.

Once you have your list of contacts, it's time to start contacting them! We recommend emailing. We also recommend making things easy on yourself and your participants by using a scheduling tool like Appointlet synced to a Google, iCal, or Outlook calendar: https://www.appointlet.com/. It's free and it will make the process so much easier if you can send a single link to potential participants so they can schedule a call with you. You can even include a Zoom meeting link in the invite that the software sends.

If the people are local, you can always meet face-to-face, as well. And as far as focus groups, that's pretty much a must, though it is possible to simply do a group Zoom call if that's more convenient or the participants aren't in the same physical location. Regardless, you'll have to use a tool like Doodle to get folks' availability to conduct focus groups (https://doodle.com/en/). Once you have time slots shared by 6–8 people, you can schedule those individuals into a focus group and send them a meeting request.

Once scheduled, it's time to conduct your interviews or focus groups. Review the tips in the sections earlier in this chapter on each method to help you prepare. Don't forget to record and take notes!

Developing Audience Personas   Once you've collected your data, the question now arises: how do you analyze it and explain this analysis to people outside of yourself? Remember that content strategists are almost always working collaboratively: with other members of organizations, with audience members, and so on. Because of this, it's important to represent audience analysis in a way that other people can easily understand.

One way to do this is by developing audience personas. A persona is an archetypal audience member, meaning it's a single audience member that represents a group of audience members. Consider how Goltz (2014) describes them:

> Each persona represents a significant portion of people in the real world and enables the designer to focus on a manageable and memorable cast of characters, instead of focusing on thousands of individuals. Personas aid designers to create different designs for different kinds of people and to design for a specific somebody, rather than a generic everybody.

The key points here are that personas represent a significant portion of people, and that they allow you to think about audience members as specific people.

We encourage people new to content strategy to think of personas as living, breathing slices of a pie chart. It's easy to gather data and put it in a pie chart, to say that investors make up 34% of your target audience, for example. But it's hard to develop content for a slice of a pie chart.

If you develop a profile for an archetypal audience member, however, with a name, a story, goals, and challenges, it becomes much easier to understand how your content can meet their needs. They become a real, specific person to you rather than just a statistic.

Analyzing Data for Patterns   Once you have all of your data from analytics, interviews, focus groups, and other sources, it's time to analyze it all. For the purposes of audience analysis, this process largely involves looking through your data in order to identify patterns that seem important to audience segments you're analyzing. The following questions will help you:

- Were there closely-related themes that arose again and again in what participants said?
- Were there important terms, images, problems, or issues that you noticed again and again?
- Were there a lot of familiar plot points in the overall story you heard or saw?
- What was distinct and different about each audience member (what didn't match up with every other audience member)?

We recommend simply making a list of patterns you notice from this initial analysis. Since the main deliverable from this analysis will be personas,

	A	B	C	D	E
1	Demographics	Stories	Goals	Challenges	How we can help
2					
3					
4					
5					
6					
7					
8					
9					
10					
11					
12					
13					
14					
15					

Figure 3.1 Example audience analysis spreadsheet

Source: Fillable version available here: https://bit.ly/3NGudkM

however, it's probably helpful to keep the following attributes in mind as you answer the preceding questions:

- Demographics (e.g., age, race, gender, location, occupation)
- Story: What makes them a good audience member for this content? What cultural values do they bring to the content?
- Goals and challenges: What is the audience member trying to accomplish with the help of this content? What pain points are they experiencing that can be alleviated through useful information?
- How I can help: What can the strategist do through their content to help the audience member achieve their goals and alleviate their pain?

Essentially, think of these different attributes as buckets that you're going to put your notes in. When you see a pattern emerge, such as a similar story that several participants shared about a type of content they really enjoy, put that in the story bucket. When you hear a challenge that many participants experienced, put that in the goals and challenge bucket.

As you'll see throughout this book, we're a big fan of spreadsheets! We recommend keeping your data analysis in one (Figure 3.1).

How Do I Create a Persona?   Once your analysis is completed, you're ready to create personas. Unless you're focused on a very specific audience segment, you'll almost certainly have more than one. How many should you have? Again, that depends on your content goals. Maybe our example nonprofit discovers that who they're really after are two types of real estate investor: one that will buy old buildings and restore them, and one that will build new buildings. So, that means two personas. Maybe they analyze further and discover that there are two types of restoration investors: people who want to buy a building and turn it into a business and people who want to divide up a large building and rent spaces. Now, they have three personas.

Your data is always the thing that will tell you where your personas lie. The patterns you've identified in your data are the building blocks of your personas. The patterns tell you if a persona represents a significant portion of your audience, or if they are an outlier. They tell you which stories are representative of an audience segment and which represent a few scattered individuals.

Once you feel like you have a sense of these patterns, it's time to turn them into clearly defined attributes that represent the entire population of your audience. As we previewed earlier, the attributes most popularly used in persona creation include:

- Name
- Photo
- Demographics (age, race, gender, location, occupation)
- Story: What makes them a good audience member for this content? What cultural values do they bring to the content?
- Goals and challenges: What is the audience member trying to accomplish with the help of this content? What pain points are they experiencing that can be alleviated through useful information?
- How I can help: What can the strategist do through their content to help the audience member achieve their goals and alleviate their pain?

Once you have filled out all the attributes for an individual persona, you end up with a deliverable like the following (Figure 3.2).

We discuss how to put personas to use when creating a content strategy plan in Chapter 7.

We discuss online templates you can use to create personas in Chapter 13, the chapter on tools and technologies. You can create a persona using any desktop publishing tool, however. The design of a persona is less important than the information it contains.

As far as *creating* an actual persona, there are two schools of thought: one we'll call *the composite method* and one we'll call *the representative method*. In the composite method, you take all your patterns from your data analysis and use them to create a fictional persona that you feel represents the entire audience segment. This method seems to be fairly popular among marketers who are constantly trying to appeal to those pie chart slices, to their core audience segments. The danger with this method, however, is that it's very tempting to create a persona who happens to match your goals, rather than the other way around. It's too easy to create a persona that wants the content you were going to create anyway, rather than someone who you have to adapt your content to.

That's why we recommend the representative method: you simply select a real, live individual from your actual data and use all of their information to fill out the persona profile. With this method, you're dealing with an actual person, so the temptation to create a persona that fits your preconceived notions is gone. The danger with this method is that you'll pick a representative who doesn't adequately represent the patterns within a key audience demographic. To mitigate this danger, once a representative individual is selected, it's important to compare this individual to the patterns you've identified in your overall data to ask yourself if they truly represent these patterns.

We end this chapter with considerations for creating a single persona. Next we discuss each attribute of a persona, in-depth.

ATTRIBUTE #1: NAME

You should give your persona a name (just a first name is fine). It shouldn't be their real first name, unless you've gotten permission from your representative individual to do so. It should also be a name that you feel is representative of them. In our example persona (Figure 3.2), Mindy's real name might have been Brittany or Bree.

Mindy

Age: 29 Years Old

Story

Although Mindy has been watching sports her whole life, she recently moved from Denver to New York and is looking for better ways to consume content for the local sports teams. She wants to prepare for the upcoming basketball season, but she knows nothing about the current state of the team(s), the players, and standings in her new, local market.

Goals or Objectives

Wants to learn about player signings and trades, the scores of the games, analysis of the best players, and the current standings. She wants to become informed about her new local teams and make a decision on which one she would like to root for.

Biggest Challenge

With several different teams playing in her new city, and different ways to access this information via apps, videos, articles, and more, she does not know which way would be the most beneficial and is overwhelmed.

How I Can Help

As a content strategist, if we need to create digestible pieces of information for the team that she wants to learn more about and allow her to customize the times in which she wants to be updated. Instead of 15-minute intervals and pushing the multimedia content (such as text, video, and tweets simultaneously), allow her to select how she would like to learn about the teams around her.

Figure 3.2 An example persona

ATTRIBUTE #2: PHOTO

Similarly, choose a photo that represents the same gender, race, age, and overall appearance of your representative audience member, but don't use a real photo of them, even if one is available, without their permission. A good solution is to choose a royalty-free, stock image from a site like *http://pixabay.com or http://www.shutterstock.com/*.

ATTRIBUTE #3: DEMOGRAPHICS

The demographics you list will, again, depend on your content goals. What makes your representative audience member important to your goals? Is it their gender? Their race? Their age? Their content preferences? Only include *pertinent* demographics, meaning the ones that matter to the context of the content you are developing or managing. Review the demographic categories in the section entitled: "A workflow for Conducting Interviews and Focus Groups."

ATTRIBUTE #4: STORY

The story is one of the most important parts of a persona. Here is where the fine-grained details of your audience segment come to life. The story should be brief but should contain details that are pertinent to your content goals. What is it that makes members of this audience segment unique? What is it that sets members of this audience segment apart from members of other segments? You don't want to go on for paragraphs and paragraphs with needless details about your persona's story. You *do* want to include the details that highlight who they are as a unique individual, and how this uniqueness affects their relationship to content.

ATTRIBUTE #5: GOALS AND CHALLENGES

While the previous attribute allows us to paint a picture as to why this audience is ideal for our content and gives some context as to how they are interacting with it, focusing on the goals and challenges of our persona is truly key to the entire audience analysis process.

We want to understand what goals our persona has in regard to our content. Are they engaging with our content in order to take action? Are they trying to accomplish some kind of objective? Do they want to learn something? The goals section should spell out why they are coming to our content in the first place.

However, we also need to understand that not everything is a positive experience. While our example persona, Mindy, may be engaging with our content on a daily basis, she may not be having an entirely positive experience with it. She may be experiencing pain points, or points in which our content fails to satisfy her needs. These pain points may include challenges that she faces engaging with our content, or the frustrations that she feels when going through our content. They can also include challenges in her own life as a consumer that she is looking to solve through our content.

By clearly describing a persona's goals and challenges, it becomes much more obvious how we can improve our content strategy to alleviate an audience's pain points and help them reach their goals.

ATTRIBUTE #6: HOW I CAN HELP

The ultimate goal of a persona is to define how we can help our audience. This section should be an articulation of how our organization's goals align with the persona's goals. How can our content help the persona to achieve their goals? How can it enable them to surmount their challenges?

This final attribute of our persona allows us to define what we as content strategists can do to help our audience.

Getting Started Guide: Conducting Your Audience Analysis Research

To get started on your audience analysis research, follow these steps as they are articulated throughout this chapter:

1. Gather any audience data you can from existing sources within an organization, including website analytics, user-generated content, and customer surveys.
2. Brainstorm potential audience segments based on demographics you think are important.
3. Recruit individuals you personally know that match the demographics of your audience segments for interviews or focus groups.
4. Conduct your interviews and focus groups. Don't forget to record, if you get permission to do so, and to take notes!
5. Analyze all your data to look for patterns across it.
6. Using this analysis, create personas that represent key audience demographics!

References

Babich, N. (2020, February 21). *The 15 rules every UX designer should know: Adobe XD ideas*. Xd Ideas. Retrieved March 31, 2022, from https://xd.adobe.com/ideas/career-tips/15-rules-every-ux-designer-know/

Babich, N. (2021, March 29). *How to conduct a user interview that actually uncovers valuable insights.* Shopify. Retrieved March 31, 2022, from www.shopify.com/partners/blog/user-interview

Court, D., Elzinga, D., Mulder, S., & Vetvik, O. J. (2009, June 1). *The consumer decision journey.* McKinsey & Company. Retrieved March 30, 2022, from www.mckinsey.com/business-functions/marketing-and-sales/our-insights/the-consumer-decision-journey

Georgiev, D. (2022, January 18). *80+ referral marketing statistics that show why it works.* Review42. Retrieved March 31, 2022, from https://review42.com/resources/referral-marketing-statistics/

Getto, G. & Labriola, J. (2016). iFixit myself: User-generated content strategy in "the free repair guide for everything." *IEEE Transactions on Professional Communication, 59*(1), 37–55.

Goltz, S. (2014, August 6). *A closer look at personas: What they are and how they work: 1.* Smashing Magazine. Retrieved April 1, 2022, from www.smashingmagazine.com/2014/08/a-closer-look-at-personas-part-1/

Ten Years on the consumer decision journey: Where are we today? McKinsey & Company. (n.d.). Retrieved March 30, 2022, from www.mckinsey.com/about-us/new-at-mckinsey-blog/ten-years-on-the-consumer-decision-journey-where-are-we-today

Thompson, G. (2016, August 17). *8 top tips for running a tip top focus group.* Bunnyfoot. Retrieved March 31, 2022, from www.bunnyfoot.com/2016/08/8-top-tips-for-running-a-tip-top-focus-group/

Further Reading

Albers, M. (2003). Multi-dimensional audience analysis for dynamic information. *Journal of Technical Writing and Communication, 33,* 263–279.

A complete guide for analyzing and defining your target audience. Point Visible. (2021, December 29). Retrieved April 1, 2022, from www.pointvisible.com/blog/complete-guide-target-audience-analysis-content-marketers/

Huddy, G. (2019). *What is audience analysis?* Brandwatch. Retrieved April 1, 2022, from www.brandwatch.com/blog/audience-analysis/

Portigal, S. (2013). *Interviewing users: How to uncover compelling insights.* Rosenfeld Media.

Walwelma, J., Sarat-St. Peter, H., & Chong, F. (Eds.). (2019). Special issue on user-generated content. *IEEE Transactions on Professional Communication, 62*(4), 315–407.

Winik, M. (2021, April 8). *What is audience analysis and why it matters.* Similarweb. Retrieved April 1, 2022, from www.similarweb.com/corp/blog/research/audience-and-brand-building/what-is-audience-analysis/

4 Identifying Content Types and Channels

Now that we've gotten a basic understanding of our audience, or to use the term from previous chapters, *the right people*, we can work on identifying different content types and channels in order to create *the right content* that addresses both audience goals and organization goals. For this chapter, we will be focusing on the identification process, which will prepare us for the following chapter on content audits.

What Is Content?

So, before we discuss different content types, it's important to define what we mean when we say "content" because it can be a vague and nebulous term. It may bring to mind the concept of matter from the sciences. Just as everything is made of matter, you might be justifiably thinking: is all communication made of content? Is a website content? Is social media content? Is a book content?

In a general sense, content is the *stuff*, the *information within a specific type of communication*, such as a website, social media feed, or book. This is still very broad, however. What's the difference between a figure, a paragraph of text, and an animated gif? Are these all just "content"?

One definition we have used for content when we are discussing the context of content strategy is *useful information an audience will see*. This is a much narrower definition than the previous one. Breaking this definition down, we can see that content must be both *information* and something *that is useful to a specific audience*. Technically, if you keep a journal or diary in which you record reflections on the day's events, then this information is content, based on the first definition: it is information within a specific type of communication. But a content strategist wouldn't really care about your reflections, because there's no external audience for them. There's just you.

Some basic questions to keep in mind when considering if something is content or not, in this narrower sense, are thus the following:

- Will an external audience see this information?
- Might an external audience use this information for some specific purpose?

DOI: 10.4324/9781003164807-4

- Does this information try to affect an external audience?
- Will this information be delivered to a specific group of people outside its creator?

Asking these questions about a piece of information should help you determine if information is content that needs to be made part of a content strategy.

Another take on defining content from Lester (2011) is "content is the presentation of information for a purpose to an audience through a channel in a form." As you can see, there are many similarities between this definition and our own. Both dictate that for information to be considered content, it must have some purpose, some use. Both definitions agree that content is information presented to an audience. This second definition, however, adds some additional details about how an audience is presented with content: "through a channel in a form." This may seem obvious. If you're seeing content, then it must be in some form. However, when trying to sort through all the information available to today's audiences, we find it useful to be exact.

This reference to content being presented through a channel and in a form is also a reference to *content types* and *content channels*, which will be defined and explored in a later section of this chapter.

Keeping this second definition in mind, we can add some additional questions to our list for determining if something is content:

- Is the information presented in a specific type or form?
- Is the information presented through a specific channel?

To help us better answer these last two questions, let's explore what we mean by content types and content channels.

What Are Content Types?

Now that we have defined content, it is time to look at content types. Throughout our daily lives, we consume a lot of content. In some cases, we may see a single type of content, such as an image of a product or a textual description of that same product. But in most cases, the content we view is *complex content*, which is a combination of more than one content type, such as a product description with an image, price, SKU, and search tags. Because of the prevalence of complex content, it is important to be able to identify and categorize content into individual content types.

A content type can be defined as "a reusable container for managing content by common structure and purpose" (Hane, 2016). Using the example from the previous paragraph, let's take a look at a product page for an online storefront to identify and categorize the different types of content present.

The example in Figure 4.1 is a product page for a catnip-filled, fish-shaped cat toy from the online storefront The Pet Beastro, which specializes in pet (dog and cat) health and wellness.

Figure 4.1 An example of content types in a web page

Source: The Pet Beastro 2022 thepetbeastro.com

https://www.thepetbeastro.com/yeowww-cat-toys-fish-bowl-stinkies-sardines-stars.html

On this product page, we have several different types of content:

- Product name—Yeowww! Cat Toys Fish Bowl Stinkies Sardines Stars Single
- Product image
- Price—$2.97 USD
- SKU—812402000720
- Weight—113
- Availability—In stock, 5
- Product tags—cat, catnip, fish, stars, madeinusa, etc.
- Product description
- Links—benefits of catnip, benefits of play, and Yeowww! products

This is just an example of the content types you can identify in a very specific channel: an online storefront. Depending on the type of organization you're working with and your intended audience, there may be a much longer list of content types, different content types, or different terminology for the content types listed.

What Are Content Channels?

Now that we have discussed content and how to break complex content into content types, we can talk about how we distribute content to an audience. This is done through *content channels*. Content channels include *any means where content is distributed in order for it to consumed by a specific audience*. This can include, but is not limited to, the following means of distribution:

- Website
- Search engine
- Social media
- Email
- Documentation
- Internal content repository

These are common channels that exist in almost every organization, to varying degrees. There are few organizations out there today that don't have a website, that aren't on any form of social media, or that don't have any types of documentation, though they do exist. And almost every organization has some kind of repository where they store all this content, though sometimes it isn't centralized or organized. And often there is more than one.

Using our example organization for this chapter, The Pet Beastro's content channels include:

- Email
- Newsletter
- Blog
- Online Storefront
- Social media (Facebook, Instagram, Twitter, Pinterest, YouTube, and LinkedIn)
- Documentation (Employee Handbook, Code of Ethics)
- Internal content repository (Website CMS)

When it comes to reaching our audience through content channels, organizations like Pet Beastro are not going to distribute every single piece of content to every single content channel. For example, text-heavy content is ideal for a newsletter or blog but would not be as effective on an image-centric social media platform like Instagram. In that case, we may decide to use an image with a short caption instead. Because of this, we need to have a *channel plan* in order to determine which content will be distributed to which content channels in order to reach our audience or audiences.

Next we talk about developing a channel plan, but first we explain why it's important to identify content types in the first place.

Why Are Content Types Important?

Here's where we start to get to the nitty-gritty of this chapter. Up until now, you may be asking yourself: ok, this is very interesting, but why is this important for content strategy? Beyond the ability to identify and categorize complex content, which is an essential part of content strategy, content types are important for tasks requiring:

- Reusability
- Consistency
- Search and classification
- Scalability

(Hane, 2016)

Essentially, content types allow you to replicate content across an entire organization's channels. If everyone in the organization is on the same page about images and how they're used in the organization's website, for example, then it makes developing image layouts on the website a lot easier. Without the ability to replicate content, you're left designing each piece of content from scratch, which takes a lot more time and effort and runs the risk of creating content that doesn't meet your content goals.

Imagine if the organization behind the aforementioned image didn't have a plan for deploying their products on their website. Image if each product used a different set of content types or used content types differently between products. Imagine if some items had product names and others didn't. Imagine if prices were missing from some products. Or imagine if prices were in different forms of currency throughout the site (pounds versus US dollars versus Canadian dollars). Imagine how difficult it would be for their website users to find a product, add it to their shopping cart, and make a purchase.

Next we delve more into this idea of replicating content across channels.

Reusability

In most cases, content will be used more than once within an organization because it was created to be viewed by an audience and to fulfill a purpose. When a new need arises, it's easiest and most efficient to repurpose existing content rather than starting from scratch. Additionally, reusing content can save time, money, and resources that are better spent on things like marketing and customer support.

Using the aforementioned product page example, this product image may get reused by Pet Beastro in an upcoming online summer catalog featuring cat toys with a star theme, or on a social media post for a daily buy-one-get-one promotion.

By breaking down content into smaller content types, the content is easier to adapt and *reuse* for future situations. For example, it is easier to reuse an image that is stored in its most basic format in future media than it is to disassemble complex content each time an image needs to be reused.

A simple example of this is if you've ever tried to use an image you found in Google in school project or work report. You may have found that the image wasn't the right size, was pixelated, or didn't match the layout of the document you were creating. This is because the image was formatted to work well on the website that Google indexed, not in your document. Now imagine if the image was stored in a universal format that can be repurposed, rather than a format designed for a specific channel. You could take the image, customize it for its new channel, and publish it, without worrying that it won't conform to the new channel because it's still formatted for the old one.

Consistency

Content types create *consistency* by allowing the same content type to be easily used for multiple purposes. Additionally, by using the same content type for multiple purposes, we can be certain that the formatting, text, and/ or visuals are consistent between channels and, ultimately, throughout an entire organization.

Using the aforementioned product page example, this particular product, the Yeowww! Cat Toys Fish Bowl Stinkies Sardines Stars Single, needs to retain this product name wherever it's mentioned. An incorrect product name in an online catalog or promotional email sent to customers could lead to confusion about the product. Additionally, a product labeled with an incorrect name at the physical storefront or in the store's inventory could lead to an incorrect stock number of the item or to the incorrect item being ordered from the supplier.

All of these issues can affect a company's credibility to its customers, employees, and suppliers. Consistency helps an organization use content effectively by ensuring that each piece of content is used correctly across channels.

Search and Classification

As we talked about earlier in the chapter, content can consist of many different types. These types get used across channels to serve distinct purposes. Let's say an employee of Pet Beastro is asked to create a new product page for a just-released cat toy. Similar to the previous example (and for consistency), all cat toys on the online storefront need to contain (and reuse) the supplemental link explaining the benefits of play. The employee will be able to complete the new product page faster if they know exactly where to find that existing link. Not having an easy way to locate the link could lead to additional time spent searching for the link or linking to the wrong content, both of which would take up additional time, money, and effort.

From the customer side, customers that are new to the Pet Beastro website might want to compare different products to weigh which ones are right for their pets. If they can't find the information they're looking for on the Pet Beastro website, they will most definitely search elsewhere online, where they may find a competitor's content, which could lead them to buy from that company instead of Pet Beastro.

And that's just one specific content type.

By breaking content down into content types, we can better organize content into different *classifications*, so it is easier to *search* for, which makes it more findable and ultimately more useful.

Scalability

Regardless of the size or age of an organization, it is important that content is organized so that it is *scalable*, meaning that the organization's content can grow as it does. Imagine an organization that gets a request for a partnership with a large supplier of retail items, such as a whole new line of all-natural cat toys. The supplier may look at the company's website and ask: can you add several hundred new pages to feature these toys within the next month?

If the organization has to format all this content manually, then their response may be: no. And in that case, they may lose out on this opportunity to a competitor. Now imagine if they have all of their products neatly organized in a central repository that allows them to quickly upload new ones to their website. This means that there is no need to change the classification names or overall structure of their online storefront as they add new products to it.

Using the aforementioned product page example, when this online storefront first opened, it's possible there were only a handful of cat toys available. In this small group of content, a category like "toys" would have sufficed to organize these products together on the company's website. The Pet Beastro online storefront now currently offers over 140 different cat toys. The category "toys" isn't very useful anymore, unless it's attached to some additional category names like "cat" or "cat nip" so that people searching for toys can narrow their search.

Now imagine that there was a limit on how many category names could be added to a specific product. This would mean that this content type wasn't scalable, because as more products get added under the same category names, those category names would become less and less useful.

Then there's also the question of how to organize content types across channels. A website is going to have different content types than a social media platform or an email newsletter. Content strategists have to manage content within many different channels, so that means knowing which content types exist in which channels.

One way to manage all that information is to create something called a channel plan, which we turn to next.

How to Develop a Channel Plan

A channel plan helps us determine the best ways to use content channels to deliver our content types to our audience. It is smaller than a content strategy plan (Chapter 6) but can help you identify which channels to include in your overall plan. Developing a channel plan will entail the following steps:

- Audience analysis
- Content and content channel analysis
- Choosing content channels
- Choosing content types
- Choosing metrics (covered in Chapter 6)
- Creating an editorial plan (covered in Chapter 6)

Audience Analysis

The first step in creating a channel plan is analyzing your audience (Dorland, 2019). If you've already read Chapter 3, you know that we recommend creating personas to stand in for groups of people within an audience. If you skipped to this chapter from Chapter 3, we strongly recommend revisiting Chapter 3 to create some personas so that you have them on hand. This will help us determine who *the right people* are and how to most effectively reach them with our content types and content channels.

People always have preferences for how to get their content. Some people get annoyed by email. Some people want a mobile app that will remind them when there's a sale. Some people just head to an organization's website when they want updates. Some people subscribe to the organization's blog. Because of this, it's important to understand who are the people you're trying to reach before you ever start talking about channels.

Once you have audience personas, however, you next need to determine which content types and content channels would be best to reach each persona. To do so, revisit the goals and challenges of each persona. Ask yourself: which content channels would best help them reach their goals and deal with their challenges?

Say one of your personas is a cat owner named Stacey who is very busy and hard to reach. Stacey has three cats, all with very different personalities. She needs information pushed to her, because she's simply not going to check a website, blog, or social media feed on a regular basis. At the same time, she is motivated to buy new cat toys on a regular basis, because her cats get bored with their toys after a few months of playing with them. She has a regular cycle of donating toys her cats don't play with anymore and shopping around for new toys that might interest them.

If this is a primary persona for your content strategy (meaning she represents a significant portion of the types of people you'd like to reach), then you need to think about the types of content that would appeal to Stacey (and thus people like her). Something like a VIP email newsletter that's triggered by purchasing behavior may be a better match for someone like this than updates to a website, blog, or social media feed.

It's not that you want to neglect the website, blog, and social media of your organization, however, unless *you don't have any audience personas that align with those channels*. See how that works? Many organizations we've consulted with think that they have to be on every content channel all the time. This isn't sustainable for most organizations, unless they have a dedicated marketing team that can sustain all that content.

Rather, it's better to think about reaching individual personas and focusing on the channels that will appeal to them. Questions like the following will help you match channels to personas:

- Which personas are currently hardest to reach?
- Which personas are currently easiest to reach?
- Which personas are the most motivated to keep up to date with your organization?
- Which ones are least motivated?
- What channels are likely to reach the largest amount of your personas?
- What channels are specific to a small group of personas or to a single persona?

The next step in creating a channel plan is to analyze what existing content types and content channels are currently available to use in your organization, or whether you need to develop new ones.

Content and Content Channel Analysis

Next, before any new content types or content channels are created, it is important to analyze what content and channels currently exist, or "what we already have" (Pulizzi, 2012). This allows us to take stock of what already exists, so that we don't duplicate efforts when creating new or additional content types and content channels.

We will discuss this process in more detail when looking at content auditing in Chapter 5. Briefly, this analysis entails determining what content (specifically, content types) and content channels already exist. These existing channels may be narrowed down to those that are owned by the organization, and what channels involve some sort of fee. The extent and detail of this analysis may be determined by your organization's priorities, deadlines, and budget.

Using the example organization for this chapter, a content type analysis would consist of searching for all content types used by the organization. Once the analysis is completed, the results can be displayed in a variety of ways. One simple way is to make a list of the different content types.

Assuming that the organization only uses the content types listed on the example product page, the list would be:

- Product names
- Product images
- Prices
- SKUs
- Weights
- Availabilities
- Product tags
- Product descriptions
- Links

Similarly, a content channel analysis would consist of searching for all content channels used by an organization. Like a content type analysis, results from this analysis can also be displayed in a variety of ways, but the simplest way is to just list all the content channels the organization currently uses. So, for our running example, this list would be:

- Email
- Newsletter
- Blog
- Online storefront
- Social media (Facebook, Instagram, Twitter, Pinterest, YouTube, and LinkedIn)
- Documentation (employee handbook, code of ethics)
- Internal content repository (website CMS)

Once you have these lists, it's time to start choosing which content channels you're going to focus on.

If you're doing this analysis as part of a run up to a full content audit (see Chapter 5), then you might want to think about putting these lists into a spreadsheet. We provide a link to an example spreadsheet in that chapter.

Choosing Content Channels

Once we know what content types and content channels we have, we can then use our audience analysis, personas, and organizational goals to determine which content channels would be best to use. This includes determining which existing content channels need to be revised or even retired, as well as which new content channels need to be added.

Using the running example for this chapter, we may determine that Pinterest has low audience interaction and is duplicating the efforts of reaching the same personas as Instagram. With this information, our organization may choose to stop using Pinterest, edit the content delivered through Pinterest, and/or consider using a new social media platform that is showing a lot of promise in reaching a certain type of person as a new content channel.

The goal is to have a clear objective for each content channel going forward that aligns with your organization's goals and needs. Additionally, having a clear and measurable objective will help determine if each content channel is successful or not.

Once you have selected the channels your strategy will focus on, it's time to select the content types that will fit into each channel.

Choosing Content Types

From here, we can determine the content types to choose for each content channel. As we already discussed, some content channels work best with text-heavy content, while other content channels require more image-based content. In truth, there are so many content types out there that it's nearly impossible to list best practices for all of them in a single chapter, or probably even a single book. In Chapter 8, we discuss the process of content modeling, which is really the process of templating out each content type and how it fits within a channel.

When you're considering choosing content types, doing a competitive analysis can help you get a rough sense of which content types fit best into which channels. There are many ways to do a competitive analysis, but one simple method is:

1. Locate 2–3 competitors for your organization (e.g., 2–3 pet store Instagram accounts or 2–3 pet store online storefronts)
2. Select one competitor you think is most worth of imitation
3. Examine 5–10 pieces of content used by the competitor (e.g., 5–10 Instagram posts or 5–10 product listings)
4. Screenshot or save links to pieces of content you think are particularly effective
5. Make notes about which content types the competitor uses and why

This quick analysis can help you brainstorm the available range of content types, but it's not definitive. To completely plan out each content type within each content channel, the process known as content modeling, you need to research more and spend significantly more time planning. But as a way to get a rough idea of what's out there, a competitive analysis can be very helpful when you're trying to identify the content types common to a specific content channel.

Also, don't be afraid to look up best practices from thought leaders during this process. Many content-focused professionals write blog posts, whitepapers, ebooks, and full-length books for their organizations. If you're not sure what an effective blog post, Instagram post, or About Us page looks like in the current marketplace, try to find 3–5 articles by leading experts on the piece of content you're trying to create. If nothing else, this will help you generate ideas.

Regardless, at this point you should brainstorm the different content types you think you'll need for each content channel. These can be very simple lists, such as:

Instagram post

- Image
- Location tag
- Hashtags
- Caption

Product listing

- Product names
- Product images
- Prices
- SKUs
- Weights
- Availabilities
- Product tags
- Product descriptions
- Links

These lists will help you later when you model each piece of content.

Choosing Metrics and Creating an Editorial Plan

Now that we have objectives for each content channel, how do we determine if each content channel is meeting objectives? The answer is metrics. Metrics allow us to measure if a content channel is meeting its objectives or not.

And what about changes that need to be made to the channel in the future? The answer here is an editorial plan, which helps us plan for updates, revisions, and changes as audience and organizational needs shift.

We prefer to put these types of information into a full content strategy plan, which is the subject of Chapter 6. It's important to have a rough idea what content channels and content types you'll be focusing on before you get to that stage, however, which is why we put this chapter first.

Preparing for the Future

As we know, the future is unknown. However, we can prepare our content so that it is easy to repurpose into future content channels. By breaking down our content into *content types*, we can be certain that our content is organized in a way that will grow and scale as the organization grows. In addition, breaking content channels down into the smaller components of content types allows for reusability within future content channels. Again, we can't prepare for everything, but these steps will make the process of creating new content easier and more efficient.

The next chapter will be about content auditing, where we will dive deeper into analyzing our current content types and content channels. There will be a general description of a content audit, which may be expanded or abbreviated based on your organization's overall goals.

Getting Started Guide: Identifying Content Channels and Content Types

If you are planning a content strategy for a specific organization, try following these steps to brainstorm content types and content channels. If you're working on content strategy in a class or other learning situation, you can simply find a website of a local organization, such as a non-profit or small business, to use as a learning case. Either way, follow these steps to identify the different content channels and content types of the organization you're focused on:

1. Locate all the existing content channels you can for the organization. These broad categories will help you get started:

   a. email
   b. newsletter
   c. blog
   d. website
   e. social media
   f. documentation
   g. internal content repository

2. Locate 2–3 competitor organizations online. These should be the same type of organization (e.g., pet store, homeless shelter, IT company). As a rule, when you're looking for competitors, you're

looking for organizations that you think are being effective. Don't waste time on competitors who you think are less effective than your own organization. Try to find leaders within your specific industry.

3. Find examples of content that match content channels you're trying to develop. You can focus on a particular competitor or look across all the competitors you've identified. You might find two blogs you really like, for example, but the third organization doesn't have one or it isn't effective. You may find an image on a website that you think would actually be effective for social media. You don't have to do a one-to-one comparison, in other words. You're really trying to find content you think is worthy of imitation.

4. After collecting your example content, identify and make a list of the different content types used in each piece of content. Remember that you may find similar, different, or more content types than those listed for the example in this chapter.

5. Collect all this information in a spreadsheet or other document for later use as you plan your content strategy.

References

Dorland, B. (2019, November 7). *The comprehensive guide to multichannel marketing planning.* DIVVY HQ. https://divvyhq.com/content-marketing/multichannel-content-planning/

Hane, C. (2016, November 16). *A useful guide to content types, part 1.* UX Booth. www.uxbooth.com/articles/a-useful-guide-to-content-types-part-1/

Lester, M. C. (2011, August 23). *What is content?* The Word Factory. https://thewordfactory.com/what-is-content/

Pulizzi, J. (2012, July 28). *7 steps to creating your content marketing channel plan.* Content Marketing Institute. https://contentmarketinginstitute.com/2012/07/creating-a-content-marketing-channel-plan/

Further Reading

A complete guide to cross-channel marketing. A Complete Guide to Cross-Channel Marketing. (n.d.). Retrieved March 21, 2022, from www.marketingevolution.com/marketing-essentials/cross-channel-marketing

Pulizzi, J. (2012, July 28). *Content marketing channel plan strategy in 7 steps.* Content Marketing Institute. Retrieved March 21, 2022, from https://contentmarketinginstitute.com/2012/07/creating-a-content-marketing-channel-plan/

Sullivan, F. C. (2021, January 5). *How to build a content & channel strategy in 2021.* Medium. https://medium.com/the-anatomy-of-marketing/how-to-build-a-content-channel-strategy-in-2021–457026ee5ac3

Thompson, S. (2020, December 21). *The importance of having a multi-channel content strategy in your marketing.* ContentCal. Retrieved March 21, 2022, from www.contentcal.com/blog/multi-channel-content-strategy/

5 Content Auditing

Now that we've discussed how to identify different content types, it's time to move on to what we do with those content types. As we mentioned in the last chapter, content types are containers for the information, the content, we consume via various channels, be they websites, social media platforms, emails, or even voice-to-text interfaces. As a culture, we generate a lot of content, a lot of information in a lot of different containers. And if no one is monitoring all these content types and how they fit into the overall content strategy of an organization, they can easily become messy, disorganized, and out of date.

You may have had the experience of getting conflicting information via a couple of different channels from the same organization. One of the authors of this book works with a local non-profit that serves the homeless community. He receives updates from them about their events via a few different channels, including email, Facebook, and their website. At one point, he was getting conflicting information about an upcoming event, one of their main fundraisers. He was invited to a Facebook event and then saw a post from the organization's Facebook page. The time and date for the event were different for each piece of content, however.

Because he had worked closely with the organization in the past, he called them to let them know about the inconsistency. After some investigation, the Lead Administrative Assistant for the organization discovered that the Facebook event was posted by a second Facebook page for the organization that they didn't even realize existed. They managed to log into the page and delete the event but were never able to discover who had posted it in the first place. Most likely one of the small army of volunteers who helps with the organization on a daily basis did it to be helpful and was too embarrassed to come forward.

The point is: this kind of issue can cause real problems for organizations. Perhaps there were donors who were planning to attend the event and then saw the two conflicting posts and changed their mind. Or perhaps they clicked "Interested" on the Facebook event but then simply forgot about the event once the post was deleted. The point is: this was far from an effective means of communicating with their audience of local community members interested in donating to non-profits and attending fundraising events.

DOI: 10.4324/9781003164807-5

If someone in the organization had analyzed all of the organization's current channels, they would have undoubtedly discovered the secondary Facebook page, merged it with the original, and the whole mishap could have been avoided. This is the primary reason to do periodic content audits: to make sure that all the content being distributed by an organization is effective.

What Is Content Auditing?

Content auditing is *the process of assessing and evaluating content for the purpose of improving it*. A content audit often precedes other steps in the content strategy process because it helps the strategist to get a sense of the state of all of an organization's content before making changes. Unless an organization is brand new and is creating content for the first time, there is probably already a lot of content in play.

As we talked about in Chapter 4, this can include lots of different types of content:

- Text
- Image
- Files
- Links
- Numbers (item or serial numbers)

And these content types often align with specific channels:

- Website
- Search engines
- Social media
- Email
- Documentation
- Internal content repositories

It's important to go in with these potential content types and channels in mind. Not every organization will use all of them, but otherwise doing a content audit can feel like searching for content in a dark room. At the same time, you don't want to begin your audit by assuming that the organization you're working with will have the same content types and channels as those listed here.

Purpose of a Content Audit

The purpose of a content audit is to evaluate and assess all the content an organization possesses, or a selected portion of it. When working as a content strategist, content is at the heart of what we do. And a lot of what we do involves reshaping and revising existing content to help it conform with organizational goals and audience-specific goals. For example, an organization may decide it wants to try to reach a new audience with its content that it hasn't reached before. Or perhaps existing audiences aren't as engaged as they could be. Usually, there is some frustration that leads to a content audit, in other words. Content audits are often used to examine the cause of messy problems with an organization's content.

There are two main activities associated with a content audit: 1) evaluation and 2) assessment. A content audit is used to evaluate existing content, meaning to *discover its value* (or lack thereof) to the organization. A content audit is also used to *assess* content, meaning to *measure its overall effectiveness* in the context of organizational goals and audience goals. A content audit needs to contain both these types of activities in order to be a sound audit. Discovering the value of content but failing to measure its overall effectiveness will make it very difficult to move forward with improving an organization's content. Moreover, you can't measure the overall effectiveness of an organization's content without first understanding its relative value for the organization.

In the example we introduced earlier, the homeless shelter didn't even realize they had a channel (i.e., a Facebook page) and that content was being distributed on it (i.e., Facebook events). This is partly because (as the author who has worked with the shelter knows) the organization has never conducted a full content audit. It's likely that they have other channels that were created at some point by a well-meaning volunteer or staff member that are sitting dormant, or worse, distributing misleading information that can cause real harm to the organization, including loss of donations that keep the shelter running.

At the same time, content audits are difficult for smaller organizations like regional non-profits to conduct. They most likely don't have staff with experience in content strategy. Or no one in the organization may have even heard of content strategy. And if they have and realize there's a problem, they probably lack the financial resources to hire a professional content strategist to conduct a content audit.

One mission behind this book is to empower people who don't know a lot about content strategy to use it as a tool to improve their organizations. Next we discuss the different parts of a content audit so that anyone new to this particular method can conduct one.

Parts of a Content Audit

The activities of evaluating and assessing are broken up into six main phases when conducting a content audit:

- Creating goals for your audit
- Conducting a content inventory
- Developing a rubric
- Assessing content

- Creating findings from an audit
- Writing a content audit report

These main phases have their own sub-tasks that ensure your content audit will be successful. Next we explore each phase of a content audit in-depth to help guide you through the process.

Creating Goals for Your Audit

The first step in your content audit is to create goals. It's important to have specific goals for your content audit so that all your activities are purpose driven. Depending on the organization you're auditing, there could be hundreds or thousands of pieces of content to examine.

There are many different types of rubrics for creating effective goals, such as SMART goals (Smart goals, 2021). Feel free to use any of these that you find useful. We prefer MAST goals, however, which stand for Measurable, Achievable, Simple, and Task-oriented. We developed this acronym specifically for working with content.

As you are learning, the work of a content strategist is very different than the work of other professionals, such as business managers, technical communicators, or marketers. We thought it was important to develop an acronym for goal setting that reflected the kinds of things content strategists find important.

Setting MAST Goals

MAST goals are goals that are measurable, achievable, simple, and task-oriented, but what does that mean in the context of content strategy?

Let's use an example from an actual past client several of the authors worked with. The was a small business with fifty or so employees that manufactured equipment for radio towers, among other technical products. We happened to connect with the organization's CEO on LinkedIn and struck up a conversation about the communication-related challenges he was facing. The CEO had heard of content strategy and was interested in learning more.

After a few additional conversations, the CEO asked if there was a way we could help him audit his website. His primary concern was that his website was not mobile-responsive. "We got penalized by Google and need to convert our content to be mobile-friendly," he said.

Their website was developed in Joomla! and was about ten years old when we started working with them. During that time, Google had begun to heavily penalize organizations whose websites didn't conform to mobile best practices, meaning websites whose content didn't display properly on a mobile device.

So, let's take this initial goal through our MAST rubric.

"Mobilegeddon" is the term many content strategists and other types of professionals (e.g., marketers, SEO specialists, e-commerce specialists) use for Google's search engine algorithm update of April 21, 2015. The update was the first time that Google officially started penalizing the search rankings of websites that didn't conform to mobile best practices. This means that websites that didn't conform to mobile best practices saw their content pushed down in search results, starting on that date. For companies trying to rank for specific keywords in order to sell products or services online, losing a coveted spot in search results could be devastating. Many businesses and other types of organizations suffered financial losses due to Google's decision.

At the same time, mobile devices have become a key technology for accessing online content, and have been for some time. At time of writing, 92.6% of the 4.66 billion global internet users access the internet via a mobile device (Johnson, 2021, September 10). With this statistic in mind, it is hard to fault Google for requiring anyone who publishes online content to ensure that it displays well on the types of devices the vast majority of internet users are using to access it.

Is It Measurable?

The first question to ask about a goal for a content audit is: is the goal measurable? This means that the goal is quantifiable, meaning *able to be expressed or measured as a quantity*. When we look at this initial goal, we can say that it is not really quantifiable. The state that the company wished to achieve for their content was "mobile-friendly," but this is not a quantity. It is a specific state but is not necessarily measurable.

Google provides specific standards for making a site mobile-friendly that are measurable, however (Gove, 2019). Some of those standards include:

- Ensuring that all calls to action are front and center on all pages
- Keeping menus short and to the point so they display well on small screens
- Making it easy to get back to the homepage
- Avoiding promotions that block users from navigating the site

You could thus quantify the goal of making a website mobile-friendly by saying something like "ensure that all pages on the website have calls to action that are in the top third of the page" or "ensure that all menus are short and display correctly on a screen that is 480 pixels wide." These are things you can count: if there are 50 pages on a website, you can check off each page that has a call-to-action that is on the top third of the page or each promotion that is within the screen view limit. The top third of the page is also something you can quantify as is 480 pixels.

Another way to quantify this goal would be to rely on Google's algorithm and to resolve all the issues that Google reported. If Google had logged 50 issues, for example, each of these issues could be systematically resolved.

Is It Achievable?

This is another question to ask about a goal for a content audit: *is the goal achievable, meaning doable given the technology and resources available?* In the case of our example goal, we can say that yes, it is achievable. Modern websites can be designed so that their content is mobile-friendly. The problem with achieving this goal, however, is that because the goal is not measurable, we won't know *when* we've achieved it. How will we know when we've achieved the state "mobile-friendly" for the organization's content? The answer is: we won't. Without a specific, quantifiable measurement system for "mobile-friendly," we won't know when we've achieved this state.

Whether or not a goal is achievable is linked closely to how measurable it is, in other words. It's impossible to know if a goal is achievable if you don't have a clear understanding of what will change when it has been achieved. You need a *quantifiable state*, or situation that is measurable, so you know when you've reached that state. Using some of the quantifiable states from our measurable criteria would help with this. We can know when all 50 Google warnings have been resolved, for example, or when all 50 pages of the website have been optimized to put calls to action on the top third of each page.

Is It Simple?

There is yet another question to ask about a goal for a content audit: *is the goal simple, meaning easily understood?* We can say that yes, our example goal is simple. At the same time, however, as we have revealed in looking at how measurable and achievable the goal is, the way the goal is stated is not *specific* enough. Simple doesn't mean overgeneralized. In this context, simple means *able to be comprehended easily.* So, though our goal is simple in the sense that we can understand it at the surface level, because it is not measurable or specific enough to be achievable, this simplicity is a false simplicity.

As we begin to dig into the goal, we realize it is not simple at all, because we don't really understand how we can achieve it. There are too many variables for a site being "mobile-friendly" for us to really comprehend what we should do to achieve this state. And as we've described earlier, the state itself is also vague and overgeneralized.

This happens a lot with initial goal setting within organizations. Organizations are often able to identify what is wrong with their content in very broad strokes, but this simplicity doesn't do justice to the complexity of the goal. At the same time, as goals get more complex, they must also remain simple enough for everyone involved to understand them, including content writers and other professionals within the organization who will be working on the content audit.

Is It Task-Oriented?

This is our final question to ask about a goal for a content audit: *is the goal task-oriented, meaning does it imply a specific action?* Content strategy involves a lot of activity. It is not a theoretical or philosophical pursuit. It is about taking action to the improve content. Content strategy goals should contain tasks that will lead to the resolution of a problem. In our running example, we can say that this goal is not task-oriented whatsoever. There are no activities mentioned that will lead to the resolution of the problem, the organization's website getting penalized by Google.

A primary activity that could help resolve this problem is, of course, a content audit. By assessing and evaluating each piece of content on the client's website, we could identify and develop a plan for revising each piece of content that is not in compliance with a Google standard for being mobile-friendly.

Revised Goal

In working with this organization to set a measurable, achievable, simple, and task-oriented goal for their content audit, we came up with the following, revised goal:

> We will conduct a content audit of all website content in order to assess its level of mobile-responsiveness, using the Google criteria for mobile-responsiveness as a rubric, and targeting specific issues that Google has already identified.

Hopefully it's clear at this point how this goal is an improvement over the original. First, it is measurable: we can quantify both the "Google criteria for mobile-responsiveness" and "specific issues that Google has already identified." We can also know when we have achieved the state implied by the goal, which is the "level of mobile-responsiveness" based on the criteria provided by Google. Again, if there are 50 pages on the website and 15 criteria, we can check off each piece of criteria for each page. Our goal is also simple, meaning understandable to anyone involved with the audit. The best way to assess simplicity is to show everyone an example goal and ask if they can grasp it. This wasn't the first proposed goal, in other words, but was developed in collaboration with the company's CEO and Marketing Manager. This final revision was understandable to them and so met the threshold for simplicity. Finally, the goal is now task-oriented because it contains the activity that is essential to resolving its implied problem. This activity is doing a content audit that assesses each piece of content against Google's criteria.

Another term for describing the extent to which a website conforms to mobile best practices is *mobile responsiveness*. The philosophy of creating websites that are built to work on mobile devices before any other considerations is called "mobile first," a term first coined by Luke Wrebloski (Wroblewski, 2011). Before mobile first, designers built websites for desktop displays and then reconfigured them for mobile. This created a variety of problems, especially as the mobile market began to explode in the mid-2000s.

A website designed primarily for display on a screen that is 1,280 or more pixels wide will arguably never perform well on a screen that is 480 pixels, or fewer. And as we enter the era of smart devices, you simply never know how small the display of your content might be!

Mobile first flipped this paradigm on its head, calling for designers to build websites, and other types of online applications, that displayed and performed well on smaller screens. This included a variety of aesthetic choices that are still with us today, from larger font sizes to simpler layouts with more white space to the famous "hamburger icon" menu that the overwhelming majority of websites now display on a mobile device in place of a full menu.

Looking for Hidden MAST Criteria and Setting Goal Boundaries

As you are developing your content audit goals, it's also important to think about hidden criteria. Having a specific goal is important, but sometimes there are other goals that are just as important to the organization or its intended audience. In the case of our running example, we asked the company's CEO *why* he was concerned with the Google penalties. This led to a conversation about the SEO (search engine optimization) of the company's website. Again we asked: why was SEO important to the CEO? And this follow-up question led to a conversation about how the company was able to attract customers for its products.

This process of digging into an organization's goals and how they intersect with audience goals is very important when creating goals for a content audit. The conversations we had with the CEO led to other goals for the content audit, for example. The main frustration of the CEO, and the ultimate purpose of our content audit, was to resolve the issues with their website in order to achieve a higher rank in Google search results. Google search results were an important channel for prospective customers looking for radio tower equipment and services. Because of the penalties from Google, the company's search engine results ranking was beginning to slip beneath those of their competitors. This was further leading to fewer and fewer prospective customers contacting them.

This context was very important for us to understand before moving forward. Understanding this context led us to add a second goal for our content audit:

> *While conducting our content audit we will also identify areas of improvement for the level of search engine optimization of each piece of website content for the audience of prospective buyers of radio tower equipment, as compared against Google's criteria for on-page optimization.*

Without this in-depth conversation, our content audit might have left out a very important organizational goal. It is hard to give a set list of criteria for locating hidden goals within organizations, but these questions might help you:

1. What are specific frustrations and pain points the organization is experiencing with their content?
2. What are problems the organization is running into when doing business processes that involve their content (i.e., marketing, sales, customer support, management)?
3. What is the overall purpose of the content audit in relation to organizational goals? What will the content audit do for the organization?
4. What is the overall purpose of the content audit in relation to audience goals? What will the content audit do for the organization's audience(s)?

Content Inventory

After creating specific goals for a content audit, the next step is to find and inventory all the content that will be evaluated and assessed. There are two types of content inventory with vastly different levels of scope. The first type is a *comprehensive content inventory*, which is an inventory of every piece of content within an organization. The second type is a *limited content inventory*, which seeks to inventory a specific subset of content within an organization.

There is no hard-and-fast rule for which type of content inventory is called for, but it's important to remember that some organizations have thousands and thousands of pieces of content. In our running example, the radio tower company had been in business for over 50 years. Our initial discussions with them began with the possibility of doing a comprehensive content inventory, but we warned them that this process would likely take weeks if not months. Even relatively small organizations generate a lot of content. For an organization that had been in business for five decades, this could equal tens of thousands of pieces of content!

Given the fact that we had already identified two specific goals for their content that focused on their website, we advised the company that we start with a limited content inventory of their website. The lesson here is that *the specific goals for the content audit should drive the type of content inventory you do.* Content inventories are time-consuming. If an organization's content goals lend themselves to a limited content inventory, we recommend starting there. You can always expand on this initial effort later. Gathering a large amount of content without a specific goal in mind can lead to a lot of confusion down the road.

Finding All Content

One of the main challenges when doing a content inventory is *finding* all the content you're trying to account for. We have never done a content audit for an organization who could accurately tell us the amount of content associated with their channels. Most organizations vastly underestimate how much content they have generated in even a single channel, such as a website or internal repository of support documentation. In our running example, the radio tower company thought that their website was only 15 or so pages, but it ended up being about 150 pages! What they thought was an audit of about 4,500 words of text ended up being an audit of almost 45,000 words! In addition, each of these pages contained numerous images, links, and other content types, all leading to a much larger content inventory than we had anticipated.

From talking to lots of content strategists over the years, this is also not uncommon. Technologies associated with content creation, including content management systems like Joomla!, content repositories like Sharepoint, and social media channels like Twitter, allow content creators to easily create and share many different types of content. What these technologies *do not* contain, however, is a system for analyzing all this content once it is published. Even adding a few pages a year to a website can cause an organization's site to balloon to dozens or even hundreds of pages. And our radio tower company is a relatively small business. We have talked to content strategists working with national brands who had websites with over 10,000 pages!

Because of all this, it's essential when doing a content audit to use some tools to help you keep track of all the content you're finding. One of those tools is the humble spreadsheet.

Using a Spreadsheet to Track Everything

When doing a content audit, a simple spreadsheet can make the difference between a neat, well-sorted inventory and a complete mess. Imagine trying to track hundreds or thousands of pieces of content by hand. A program like Excel or Google Sheets can be a lifesaver as these programs allow you to create a simple database that sorts all the content you're finding.

Use headings like the following to organize your content inventory into discrete categories to get started on your analysis:

- Content source
- Content type(s)
- Content channel
- Date published
- Assessment criteria #1
- Assessment criteria #2
- Etc.

> If you're looking for a handy template to get started with your first content audit, we've created a free, downloadable Google Sheet (Figure 5.1).
>
> To get it, go to the link below, click File, and then click Download. You can download the template as an Excel document, an OpenDocument, or a CSV (comma-separated values). If you're signed into your Google account, you can also go to File > Make a Copy and create a Google Sheet version that you can edit directly that way.

First, the content source should be a link to the content itself. If this content is housed within a public webpage, then it's easy to link to. If it's a physical document, the best thing to do is to scan a copy of the document and store that scanned copy in a place that it can be linked to. Some types of repositories, such as Google Drive, allow content to be linked to, and others

	A	B	C	D	E	F
1	Content Source	Content Type	Content Channel	Date Published	Assessment Criteria #1	Assessment Criteria #2
2						
3						
4						
5						
6						
7						
8						
9						
10						
11						
12						
13						
14						
15						
16						

Figure 5.1 Example content audit spreadsheet

Source: Fillable version available here: https://bit.ly/30jUQaT

don't. Regardless, it's essential that your inventory links to the actual content you're inventorying or you will quickly lose its specific location.

Next, content type(s) should describe the specific types of content contained. This can be tricky, as many individual pieces of content (such as a webpage or user manual) contain multiple content types (e.g., image, text, link). Be as descriptive as possible here, but also focus on the types of content you're examining. In our running example, since our assessment criteria focused on Google's criteria for mobile-friendliness, we inventoried each content type that could affect that criteria. This includes images, text, and calls to action (i.e., buttons that ask users to navigate elsewhere).

The content channel column should indicate the place the content is published to, which may include:

- Website
- Search engines
- Social media
- Email
- Documentation
- Internal content repositories

Again, being specific is important here. Is there one website being inventoried or multiple? Is there a specific search engine being assessed or more than one? The more descriptive you can be here, the better.

Date published will not always be easily available. Some content technologies, such as WordPress, keep track of the last update to each piece of content. Some do not. Don't worry if this information isn't easily accessible. Just leave this blank or indicate "not apparent" if it isn't immediately obvious.

As far as assessment criteria, this is just a note to let you know that you will want to include them in your overall spreadsheet, which will help in your analysis. It's okay if you don't have your rubric fully developed once you begin your inventory, however, as long as you have specific goals identified. You can develop your assessment criteria after you complete your inventory.

Using a Website Crawler to Find Webpages

In order to help you with your work, you should absolutely use a website crawler if any of the content you're assessing is attached to a public website. A website crawler is simply a tool that can find every webpage that is associated with a specific domain (e.g., Google.com).

Nowadays, many website crawlers are attached to SEO tools, such as:

- *http://www.screamingfrog.co.uk/seo-spider/*
- *https://moz.com/*
- *http://www.sureoak.com/seo-tools/seo-website-crawler-tool*

There are more here as well:

- *http://www.octoparse.com/blog/top-20-web-crawling-tools-for-extracting-web-data*

Some of these website crawlers allow for free crawling of a certain amount of links and then require payment. Some of them require payment up front to crawl any website. Regardless, you will save yourself hours and hours of time by using one of these tools if you are inventorying online content. These tools also solve the problem of finding webpages that are attached to a specific domain but not included in navigation links. These webpages can be almost impossible to find without crawling the website as they are not actively linked to any specific menu on the organization's website.

The other advantage of these crawlers is that sometimes they can help you identify difficult-to-ascertain issues such as SEO problems. Search engine algorithms change constantly, so staying on top of their criteria for ranking content can be difficult even for SEO experts. We recommend relying on a tool like one of the aforementioned if you're doing an audit that focuses on SEO.

Developing a Rubric

As far as the rubric for assessing your content, either before you start your inventory or after (it really doesn't matter—go with whatever makes the most sense to you), you'll need to develop formal criteria for *assessing* the content you've inventoried. We've already done a big chunk of the *evaluation* of this content by setting specific goals with the organization. This answers the question of what content they value. Now we need to *measure* each piece of content against that criteria.

Turning Goals into Criteria

When developing your rubric, it's thus essential to revisit your goals, which should contain measurable, achievable criteria for doing your audit. In our running example, the goals we identified for our content audit with the radio tower company were:

Goal 1. *We will conduct a content audit of all website content in order to assess its level of mobile-responsiveness, using the Google criteria for mobile-responsiveness as a rubric, and targeting specific issues that Google has already identified.*

Goal 2. *While conducting our content audit we will also identify areas of improvement for the level of search engine optimization of each piece of website content for the audience of prospective buyers of radio tower equipment, as compared against Google's criteria for on-page optimization.*

From your content audit goals, you'll want to extract simple, understandable criteria for each goal. The point of doing this is so that you can quickly and easily measure each piece of content you have inventoried. This means you need to capture each criterion from your goals and list it as a heading in your spreadsheet.

These are the criteria we identified from the aforementioned goals:

- Content meets all standards for mobile best practices (name any standards it doesn't meet)
- Content has a specific Google warning associated with it (name warning or warnings)
- Content is optimized for a target keyword
- Content meets SEO on-site best practices for target keywords (name any best practices it doesn't meet)

The most important things about these criteria are that they 1) have simple, yes/no answers and 2) allow for specific elaborations. You don't want to have to revisit a piece of content a second time once you assess it. At the same time, you need to be able to quickly scan your spreadsheet so that you can tell how each piece of content measures up against your criteria. Depending on how you like to organize information, you might want two columns for each criteria: a column for yes/no and a column to add notes on the specific attribute you're assessing (e.g., a specific Google warning).

Assessing Content

Once you have conducted your content inventory and developed your content rubric, it's time to do your assessment! Again, this should be a simple, checklist-like process at this stage. Once you have your criteria clearly spelled out in the columns of your spreadsheet, you simply go through each piece of content and assess it. Sometimes, issues can arise as you do so, however, including difficulty with ascertaining whether a given piece of content meets a given criteria or not.

Issues with Assessing Content via Your Criteria

There are several issues that can arise when assessing content. The first of which is that as you are going through your content, you may find that you're having difficulty ascertaining whether a given piece of content matches a given criterion.

If you run into this issue, ask yourself the following questions:

- Is the criterion clear?
- Is the criterion measurable?

- Can you *see* evidence of the criterion through a visual examination?
- Do you need a tool to help you see evidence of the criterion?

First, you may find that your criterion is not clear. If so, revisit your goals and ask yourself if they are truly measurable and achievable. If not, specify them, revise the specific criterion giving you trouble, and try again. Repeat as necessary until you are completely clear what you're looking for and can find evidence of it when looking at your content.

You might also find that you simply can't *see* evidence of the criterion through a visual examination. If you're doing an SEO analysis, for example, you won't be able to see something like a meta-description without knowing where to look or without the use of a tool. See more details on SEO in our next section on different types of assessments.

Different Types of Assessment Criteria

There are many different types of assessment criteria available when conducting a content audit. They range from the very simple and quantitative to the very complex and qualitative. Every type of assessment is defined by the *goals of the audit,* which are in turn driven by the *organization's content goals (organizational goals + audience goals).* Below are some of the more common criteria we've encountered when doing content audits and when researching best practices associated with them.

> We provide a workflow for assessing content at the end of this chapter as part of the discussion guide called Doing Your First Content Audit. This discussion guide will walk you through the entire process of creating a content audit in simple steps, starting with identifying and inventorying content and continuing through the process of assessing content, analyzing results, and reporting your results back to an organization.

Numerical

The first type of criteria is simple counting. It is often useful to deliver counts of different content types or sources from a content audit. This can be useful for calculating percentages later. For example, an organization may be trying to assess what percentage of its content meets a certain criterion, such as audience appropriateness. It is relatively easy to calculate these percentages if you track all of your content and associated criteria in your spreadsheet. Having a percentage also tells the organization how much work it will be to revise their content. Is 50% of their content out-of-whack? Or is it 90%? Giving simple numbers can also help organizations who are unfamiliar with content strategy to better understand your findings.

For various reasons, you might also decide to score content on a scale. If you are assessing primarily qualitative criteria, such as credibility, you might develop a scale from 1 to 10 for how credible the content is, given specific factors. This can also help you deliver an easy-to-quantify report to the organization you're auditing. It's a lot simpler to say that their content, on average, scores 6/10 on credibility than to say something like "it's mostly credible."

Website Analytics-Based

Other relatively quantitative measurements of content are criteria that are based on website analytics. Website analytics are measurements of user interactions with online content and include:

- The webpages that are most visited within a website
- The length of time a user spends on a specific webpage
- The bounce rate, or what percentage of users navigate to a webpage and then immediately abandon it

---

We provide a screenshot of an example website analytics dashboard from Google Analytics, one of the most popular analytics tools, in Chapter 2.

For a deep dive into website analytics, we highly recommend this course on Google Analytics, one of the most popular tools for measuring website analytics: *https://analytics.google.com/analytics/academy/course/6*

These measurements can be useful if you're attempting to understand an audience's goals for navigating online content. What many organizations think of as their best, or most popular, content, for example, is often not the content favored by their audience members. And there are often large portions of online content that hardly ever gets visited. Knowing what content audience members are actually visiting and how long they're dwelling with that content can tell you what content they're most interested in.

SEO-Based

SEO can be defined as a measurement of *how optimized online content is for the criteria that search engines use to rank their search results*. SEO is very complex but is growing in importance as more and more organizations seek to reach their primary audiences through search engines. Google is by far the most important search engine for optimizing online content as, at the time of writing, it represents 92% of all search traffic (Johnson, 2021, October 8).

Google's algorithm also uses several hundred criteria to rank content. Because of this, we highly recommend relying on tools to assess content for SEO. We mentioned several of them earlier in this chapter when discussing website crawlers, which are often bundled with SEO tools.

At the same time, if you're optimizing content for SEO, it's also essential to understand what SEO is and how it works so that you can make intelligent assessment decisions. For learning about SEO, we highly recommend Moz's continually updated introduction to it, The Beginner's Guide to SEO (*The beginner's guide*).

Some of the most common SEO assessments include:

- Keyword research: determining which phrases are most popular when audiences search for the type of content the organization is hosting (e.g., keywords associated with radio towers)
- Keyword optimization: making sure that content is focused on a particular keyword phrase that users enter into search engines in order to find it
- Keyword placement: making sure that this phrase is placed strategically throughout the content

The Beginner's Guide to SEO is the best place to learn how all these types of assessment affect SEO. We highly recommend reading at least the first few chapters before trying to do an SEO assessment.

Readability

Readability is a measurement of *how easy it is for a specific audience to comprehend textual content*. It can be measured quantitatively through two scales known as the Flesch Reading Ease scale and the Flesch-Kincaid Grade Level scale (*Flesch reading ease*). Readability is very important when targeting an audience that has significantly less expertise in a subject matter than the person writing the content.

Readable is an online tool that will assess a certain amount of content for free: *https://readable.com/text/*. We highly recommend using a tool like this to assess content for readability rather than doing it manually unless you are very well versed in how to score content for readability.

Usability

Assessing content for usability involves investigating if content can be easily used by a target audience member for specific tasks. In our running example for this chapter, we could have assessed the client's content for how easy it would be for a website visitor to learn more information about their products and services, or to sign up for their newsletter. We discuss usability and how to assess content for it in Chapter 10, so visit that chapter to help you if you want to do this kind of assessment.

Accessibility

Assessing content for accessibility involves investigating if content can be easily used for specific tasks by a target audience member who has a disability. In our running example for this chapter, we could have assessed the client's content for how easy it would be for a website visitor who is blind to learn more information about their products and services, or how easy it would be for a website visitor who has a mental or physical impairment to sign up for their newsletter. We discuss accessibility and how to assess content for it in Chapter 10, so visit that chapter to help you if you want to do this kind of assessment.

Authoritativeness

Authoritativeness is a measurement of the extent to which a piece of content is acknowledged for accuracy and excellence. This matters the most when an organization's content is going to be compared to content from another organization, such as a competitor in the same marketplace. In our running example, our radio tower equipment company wanted to project the image that they are a leader in the types of technology that they sell. If they wrote a blog post that described a certain type of equipment installation and it was later discovered that a competitor had come up with a cheaper, simpler installation process, then the company could lose business.

One of the best ways to assess content for authoritativeness is to recruit subject matter experts (SMEs) to review the content. A radio tower engineer whose job entails developing processes for installation of equipment would be a good source for this type of information, for example. The engineer could be asked to review a piece of content to give feedback on how accurate it is. Typically, companies must pay subject matter experts for this type of service, unless they are employees of the company itself, but it is well worth the investment to ensure the company's content is both accurate and excellent.

Credibility

Credibility is a measurement of how trustworthy a piece of content is. In the previous example, if a customer followed the procedure for installing their newly purchased radio tower equipment and found that it didn't work at all, they might decide that the company was not trustworthy. They might in turn question if the equipment they purchased was the best available and may even return it in order to purchase the same product from a different supplier.

Like authoritativeness, judging the credibility of content requires some subject matter expertise in the topic. If the content strategist isn't an expert on the topic of the content, then they won't be able to easily judge whether it is trustworthy or not. Again, they should recruit a subject matter expert to review the content in order to judge whether or not it is trustworthy.

In Chapter 7, we describe a process for interviewing subject matter experts in order to assess the authoritativeness and credibility of content.

Relevance

Relevance is a measurement of how closely connected or appropriate content is to its goal. Remember that content goals can be defined as a combination of organizational goals and audience goals. So, if an organization's goal for a piece of content is to explain a process (e.g., how to install a piece of equipment in order to support their customers after purchase) and an audience's goal for a piece of content is to learn how to complete a process (e.g., how to install a piece of equipment they've just purchased), then the content strategist can assess how appropriate the content is for those goals.

Assessing for relevance is relatively simple, then: simply articulate the goal of the content (organizational goal + audience goal) and then assess how relevant the content is to that goal. This is best done as a scale. For example, 1 can be defined as "Not at all relevant" and 7 can be defined as "Very relevant." It can also be useful to make notes on why the content was scored the way it was. Perhaps, for example, there is a lot of extraneous information that dilutes the goal of the content. Or perhaps there is not enough information to achieve the goal.

Visual Appeal

Visual appeal is a measurement of how well a piece of content adopts aesthetics that appeal to its target audience. Visual appeal is typically measured from a design standpoint, using principles of good design. One of the best reference books for people new to design is the book *White space is not your enemy: A beginner's guide to communicating visually through graphic, web & multimedia design.*

In order to assess content for visual appeal, the content strategist must have some grounding in effective design. Without it, he or she will simply be applying his or her biases to the content. We all prefer certain colors over others, for example, but it's another matter to assess whether the use of color on a webpage or book cover is *effective*, from a design standpoint.

Some of the elements of visual appeal that can be assessed during a content audit include:

- Layout: How effectively are all the elements within the content laid out as a whole? Is the layout logical? Does it follow a clear organizational scheme (i.e., alphabetical, hierarchical, chronological)?
- Color scheme: Does the content use color effectively? Does the use of color follow a discernible pattern, such as the organization's approved color palette? Does the use of color provide sufficient contrast between elements?
- Use of white space: Is the content too crowded? Too sparse? Does it use sufficient space between elements without leaving spaces that could be better utilized?
- Use of images: How appropriate are the images to the content's goals? What is the tone of the images? Are they meant to be friendly? Funny? Serious? Professional? Scholarly? Kid-friendly? How well do they match the intended tone?

Interaction Design

Interaction design is a term that can mean the ways in which a piece of content allows for interaction from its audience. The interaction design of a piece of content displayed on a mobile device will be very different from the interaction design of content displayed on a large overhead display with a separate touchscreen interface. Interaction design is typically measured through an inventory of all the ways an audience can interact with a piece of content, followed by an assessment as to whether these types of interaction are appropriate, given the goals of the content's target audience(s).

Another way to think about the interaction design of content is to ask *what actions an audience wants to take in response to content*. So, for example, if an audience needs to be able to search for content via a well-organized database, then the ability to take that action needs to be made available to audience members. On the other hand, if a search feature is included in the interaction design of content, but searching isn't an action that audience members need to take to achieve their goals, then this feature may confuse users or get in the way of actions they want to take, actions like subscribing, sharing, and posting comments.

Some of the elements of interaction design that can be assessed during a content audit include:

- Audience actions: It's often useful to simply inventory all the actions that a piece of content allows users to take. Then, the content strategist can ask for each action: does this align with a goal the audience has? There are many, many actions that audiences can take in response to content, including reading, writing, subscribing, sharing, posting comments, searching, sorting, saving, adding an item to a shopping cart, and so on.

- Calls to Action (CTAs): Calls to Action or CTAs are the ways in which content *signals* to an audience that an action is available. A brightly-colored button with the word "subscribe" indicates that clicking on this button will allow the user to receive future updates on this content. Best practices for CTAs include that they should be simple, that they should fit the audience's understanding of the action they make available, and that they should be a subtle command (e.g., "subscribe" not "would you like to subscribe?"). A lot of people think of CTAs from a marketing standpoint, meaning how persuasive they are to an audience, but this is a slightly different concern than what we're suggesting, which is that content strategists inventory all available CTAs within content and then assess if the actions they make available are simple, understandable, and match an audience's goals for the actions they want to take.
- Usability: Along with assessing the actions content makes available to an audience and the ways in which the content signals these actions through CTAs, it's also useful to assess the usability of these actions. Content strategists can ask themselves if these actions are easy to take for a target audience, for example. They can ask themselves if the actions are learnable. They can ask themselves if the actions are efficient.

Interaction design is not just a term for describing how content is designed to allow audiences to interact with it, it is also a discipline in its own right. According to the Interaction Design Foundation:

> Interaction design can be understood in simple (but not simplified) terms: it is the design of the interaction between users and products. Most often when people talk about interaction design, the products tend to be software products like apps or websites. The goal of interaction design is to create products that enable the user to achieve their objective(s) in the best way possible.
>
> (Siang, n.d.)

Interaction design thus has many elements that go beyond content. It is often the province of UX designers, interaction designers, and other types of professionals responsible for building digital products and services.

   We advise including *elements* of interaction design when auditing content because many forms of content are now highly interactive. Even if the textual information in your content is well written and appropriate for its audience, if the actions the audience expects to take with this type of information (e.g., subscribe to a blog to receive future updates) are missing or poorly crafted, then the content won't be successful at helping audiences reach their goals. Because of this, many content strategists work closely with designers to develop and manage content in a way that meets design standards for the type of technology the content will be delivered via (e.g., a mobile app versus a website versus a large outdoor display).

Audience Appropriateness

Audience appropriateness is a measurement of how well a piece of content meets the needs of target audience(s). Audience appropriateness is typically measured from a qualitative standpoint, meaning that a content strategist looks at a piece of content overall, reviews the goals of target audience(s), and then asks themselves how well the content's goals align with those of the audience(s).

In order to assess content for audience appropriateness, the content strategist must have a firm understanding of the goals of target audience(s). Once he or she understands these goals, he or she can review pieces of content and ask him or herself where the content fulfills these goals and where it falls short.

Some of the elements of audience appropriateness that can be assessed during a content audit include:

- Tone: How well does the content adopt a general attitude that would appeal to target audience(s)?
- Purpose: What is the overall purpose of the content? To inform? Persuade? Educate? Entertain? How well does this purpose align with the reasons audience(s) are approaching the content? Where does its purpose not align with audience expectations?
- Genre: What type of content is this? A blog? An article? A chapter? A webpage? A help doc? Does the purpose of this type of genre align with audience expectations? Meaning, for example, if the audience is looking for help, is this type of content where they'd expect to find help?
- Style: What conventions does the content adopt and do these conventions align with audience expectations? By conventions, we mean the ways in this type of content is *usually composed*. Here style and genre are very much aligned. Audiences recognize different content types because they have encountered similar content in the past. Not including the stylistic elements commonly associated with a specific genre (e.g., a table of contents in a book including page numbers where each chapter starts) can confuse audiences or even alienate them. Another

way to think of style is as *all the elements or basic components of a genre* (e.g., a blog post will have a title, a body, a conclusion, an image, and will be written for a non-specialist audience). Regardless, the best way to identify appropriate style is often to review examples of the genre. If creating a new help forum for the first time, for example, a content strategist might review the help forums of competitors to identify what is successful and what isn't.

Creating Findings From an Audit

Once you have completed your content audit, it's important to explain your findings to stakeholders. Whether you are working inside an organization as a full-time content strategist, working as an external consultant, or doing volunteer content strategy for a local non-profit as part of a service-learning course, you can't just deliver a spreadsheet full of your notes to stakeholders and expect them to understand what to do next.

It's important that you analyze the notes in your spreadsheet to look for patterns and then explain next steps to your stakeholders. After all, the point of an audit is not simply to find issues with content, but to identify how these issues can be fixed.

Typically, content strategists will do this in one or more of the following forms:

- A detailed report of findings, including patterns you noticed across your audit and recommendations for next steps
- A PowerPoint, Google Slide, or other type of visual presentation that leads stakeholders through findings
- A content strategy plan (see Chapter 6)
- An editorial calendar, or visual workflow, that presents a future plan for developing, editing, and managing content

In order to create deliverables like these, the first step is to rigorously analyze the data from your audit.

Analyzing Your Audit

Analyzing the data from your audit can take several forms, but we recommend a few different approaches:

1. A quantitative, percentage-based approach that matches your inventory to your specific findings
2. A qualitative approach that describes patterns regarding specific content attributes (e.g., audience appropriateness)
3. A recommendation-based approach that describes concrete actions your stakeholders can take to improve their content

The Quantitative Analysis

First off, as you review your audit notes for patterns, we recommend you keep track of how many pieces of content follow a specific pattern. As you review your notes on typography, for example, it's useful to keep a numerical count of how many pieces of content include a font that appears two small on a desktop screen, or a font that seems inappropriate for the type of audience the content is catered toward.

You can do this in several ways:

- Note a pattern in a separate document and then record each cell from your spreadsheet that corresponds to that pattern (e.g., F35).
- Color code each pattern in your spreadsheet as you encounter it so that you can easily count each instance of a specific color later. Be sure you create a key somewhere so that you know which color corresponds to which pattern!
- Do a simple keyword search in your spreadsheet document to note every instance of a specific word or phrase. Note: this will only work if you were completely consistent in your notes in the way that you described specific problems.

Don't sweat it if you are a little bit inconsistent in your counts. You're not doing statistical analysis here. This count is useful because it can indicate to stakeholders *how big a problem* a certain issue is. If you noticed ineffective CTAs in 9% of content, this is a much easier problem to fix than if you noticed it in 90% of content. You're not trying to be scientific here. You're trying to help your stakeholders understand roughly how much of their content has a specific issue.

Remember that your content audit spreadsheet is always your most important document, especially as you work with a stakeholder to actually fix the issues you uncovered. If your stakeholder agrees that the layout of many of their webpages is ineffective, for example, you're going to want to work with them to fix the ineffective layout on *every page in which it appears*. In this way, you *do* want to deliver your spreadsheet to your stakeholders along with your report so they can look at the data for themselves.

One minor note of warning: keep in mind that your audit spreadsheet is a stakeholder-facing document, so don't let your frustration with your content audit, which can indeed be very tedious, spill over into it. Two of the authors of this book were once working with an intern on a content audit for a client and the intern wrote in one cell that the client's color scheme "sucks." They went on to describe why but obviously this was not an appropriate way to report this finding to the client! Luckily, the project manager caught this before the spreadsheet was sent to the client along with their final report, but this could have easily slipped through and would have been very embarrassing for everyone involved if the client had seen it.

The Qualitative Analysis

The core of your content audit analysis is qualitative. The goal of any content audit is to explain to stakeholders the patterns you noticed across their content regarding the assessment criteria you decided upon before beginning your content. Content is never developed in a vacuum. If stakeholders decided upon a particular style of images for use in their WordPress-based content management system, for example, likely that style of images was used throughout the system.

The real goal of identifying patterns within your content audit notes, then, is to be able to point out *systemic* issues in your stakeholders' content. You don't want to completely ignore a single instance of an issue, such as a single grammar issue you noticed in one paragraph of a blog post, but repairing ineffective content is a lot of work. You'll definitely want to prioritize the largest issues first. Don't ignore that single grammar issue as you work with your client to improve their content, but if Google has removed 85% of the client's website content because it is not mobile-responsive, this is the issue to focus on!

Without noting patterns from your audit, it's nearly impossible to write a report on your findings, which should summarize your notes. Mentioning a pattern in your report signals to your stakeholders that this issue is important and should be prioritized. If you focus on issues that don't represent large samples of your client's content, then your client will come away from your report with a skewed understanding of the state of their content.

To note patterns in your audit notes, follow this simple procedure:

1. Read back through your content audit spreadsheet and use one or more of the techniques from the quantitative analysis section to note each instance of a particular issue.
2. Make a note of the most common or pervasive issues (e.g., content that is no longer relevant to its target audience) in a bulleted list somewhere.
3. As you complete your quantitative count of issues, review your bulleted list to ensure that it contains the largest percentage-issues (e.g., 60% of content was deemed no longer relevant to its target audience). You can then either remove issues that represented a much smaller percentage of your notes (e.g., 3% of content was deemed to be outdated) or note them in a miscellaneous category. Again, the point is not to ignore them, but to focus on the biggest issues you noticed in your report.

Checking Your Analysis Against Your Data

Part of being a good researcher of any kind means verifying that your findings are representative of your total data set. In the case of a content audit, you simply want to ensure that the patterns you have noted are *representative* of your notes. If you say that 58% of your client's content doesn't follow a specific SEO guideline, and they look at your spreadsheet and realize it's actually only 15%, you will lose credibility in the eyes of your stakeholders.

Again, however, you don't have to make scientific claims about your content audit. If you miscount and it turns out that only 56% of your client's content follows a particular pattern, but in your report you stated 58%, this is not a huge concern. You want to be careful in your counting, but the goal of your report is simply to indicate which patterns represented the largest samples of the client's content, so they understand which issues are most pressing.

To reiterate: your spreadsheet is the gold standard for what you found as it represents a one-for-one accounting of each issue you detected within your assessment criteria. Over many years of doing content audits, none of the authors of this book have ever had a client challenge us on the specific percentages we indicated in a report. And the vast majority of them deeply appreciated the percentages as it helped them grasp the magnitude of specific issues.

Draft Recommendations

One of the most important moves to make in your content audit report is making recommendations. Depending on the status of your client's content, there could be dozens or even hundreds of individual issues with their content. Content strategists working for large companies frequently deal with thousands of pages of content during their audits. Imagine how many issues that would entail!

That's why it's important to note possible ways to solve the issues you're uncovering in your audit as you analyze your notes. There are a few ways to do this:

1. Create a new column in your spreadsheet entitled "recommendations" or something similar and note any possible solutions that come to mind as you're doing your analysis. Note: we recommend taking this column *out* before you submit your spreadsheet to your client, however, as otherwise it may lock you into these draft solutions. You may find better solutions after you deliver your report, in other words, and don't want to be tied to your initial impulse.
2. Create a bulleted list in another document of possible recommendations. Be sure you indicate which solutions you are proposing for which issues, so you don't get confused later when you try to write your report.

Once you've completed your analysis of your content audit notes, it's time to draft your report, a process which we turn to next.

Writing a Content Audit Report

After you've completed your audit and analyzed your notes from your audit, you should have more than enough material to write a detailed report of what you found. There are many different ways to write a content audit report, but we've used the following structure in our teaching, and even with our own clients, and have found it to be a successful way to convey findings:

1. **Executive Summary:** Provide a brief, 1–2 paragraph summary of your findings from your research including:
   a. What the overall purpose of your content audit was
   b. Any trends or patterns you will be highlighting from your content audit
   c. Recommendations you are making based on your content audit
2. **Findings:** Describe your findings from your content audit in detail, being sure to use examples such as screenshots, quotes, or links to illustrate your main points.
3. **Takeaway:** Tell your audience how the website (or other channel) you audited can be improved. What are concrete actions the audience can take to improve the website?

A common question people new to content strategy ask is: how long should your report be? Our answer is always that it should be long enough to thoroughly detail all the patterns you identified in your analysis of your content audit notes. This means it can be as short as a few pages. When dealing with longer audits, however, we've written reports that easily filled up 10–15 pages. The important thing is to:

1. Document all the patterns you found in your analysis
2. Provide a recommendation for each pattern

Next, we consider matters of audience for the report.

Who Are You Writing For?

An important question to ask yourself is: who is this report for? Though this framework has served us, and our students, well for dozens of different types of clients, ranging from small non-profits to large corporations, all of these audiences had different needs. To draw on the example we used earlier in this chapter, the CEO of the radio tower equipment company was very well versed in the issues with mobile responsiveness Google had identified. He understood what a Call-to-Action was and why it was important for all content to conform to smaller screens. He understood what the Google algorithm was and that it was responsible for how his website content was ranked in search results.

Our report to him was thus able to use terms like "the Google algorithm," "mobile responsiveness," and "search results ranking." From previous conversations with him, we gleaned that he would understand these terms. Many of our service-learning students have worked to develop content strategies for small non-profits, however, whose directors *don't* understand terms of this nature.

We're definitely not advising you to "dumb down" your report by not including technical information. It's important that a client who asked for an SEO audit *receive information on SEO in their report*. What we are saying is that you need to cater your report to your audience. If your audience doesn't have expertise in terms related to the goals of your audit, then you'll need to define all those terms for them throughout your report. One of the most important purposes of a content audit report, after all, is to *educate* stakeholders regarding why their content is ineffective. This involves explaining to them the underlying principles of content strategy that we explore throughout this book.

You also need to ensure that your audience *understands* this explanation by describing the patterns you identified in terms familiar to them. When in doubt, we tell people new to content strategy to think about explaining their findings to a family member who isn't too good with technology. If your father or grandmother can understand what you're saying, chances are your client will be able to understand as well.

Responding to the Goals of the Audit

It's also important in your report to remind stakeholders what the goals of your audit were. The easiest way to do this is to make your goals into headings in the findings section of your report. You can even label them like the following:

1. Google Criteria for Mobile-Responsiveness #1: Homepage and Site Navigation
2. Google Criteria for Mobile-Responsiveness #2: Calls to Action
3. Google Criteria for Mobile-Responsiveness #3: Menu Structure

Breaking your goals down into headings in your findings section will help you cover all the parts of your initial goals in a structured way. It will also help your stakeholders understand how the report fits into your overall audit.

Documenting Findings

Within your findings section itself, it's very important to document what you found in your audit. By document, we mean *show* or *demonstrate*. It is one thing to say to your stakeholder audience that their website's homepage and site navigation doesn't conform to Google best practices. You can even lay out what those best practices are. But remember: your audience probably doesn't know what it means for a website to "[f]ocus your mobile homepage on connecting users to the content they're looking for" or to "[m]ake secondary tasks available through menus or 'below the fold'" (Gove, 2019). If your audience understood the best practices behind your audit goals, they probably would have already implemented them!

So, besides *reminding* your audience of the goals of your audit and *explaining* which best practices their communication genre failed to follow, you also need to *show* what this means. To do so, we recommend using screenshots for any communication genre that isn't primarily text-based. A screenshot showing an overly complicated menu structure or an image that isn't clear enough for its intended audience is a lot easier for an audience unfamiliar with a particular genre to grasp than a written description alone. Of course, your written description of the problem should explain what is going on in the screenshot.

You certainly don't need screenshots for every instance of a problem, either. Rather, choose a single screenshot that can represent a *category of problems*. Say there are five different places within a menu structure in a website that are too complicated for mobile navigation. And say three of them are too complicated because they have more than one dropdown sub-menu and two of them are too complicated because their header text is too long to fit comfortably on a mobile screen. In this case, you'd want to include two screenshots: one showing an example of the dropdown issue and one showing an example of the overly long text.

We *do* recommend a one-for-one relationship between screenshots and types of problems. If you have fifteen broad categories of issues that you discover in your audit, and each of these broad categories represents three-to-five specific issues, then you'll want to include thirty to forty-five screenshots in your report. This may sound like a lot of work, but trust us: it is far less work than trying to explain to a frustrated client what all the terms you used in your report mean.

Making Concrete Recommendations

Finally, it's important that you make concrete recommendations in your report for improving the content you audited. This section of your report, however, can vary strongly depending on whether you'll be involved directly in repairing the issues you found. If you are working as a consultant for an organization and that organization will use internal staff to fix the issues you've found, then your recommendations need to be very detailed. You essentially need to explain how to fix each and every issue you detected.

If you will be working to fix the issues for the organization, however, then your recommendations can be more brief and can simply lay out the type of work you will be doing to repair each category of issue. It's probably sufficient to say that the menu structure of a website needs to be simplified by removing secondary dropdown menus in favor of linking these pages within the website itself if you will be the one repairing the website's menu. On the other hand, if you will *not* be responsible for repairing the menu, then you'll need to explain to your audience *how exactly* the menu can be repaired, such as what pages in which you'd recommend incorporating the tertiary links.

If you're not going to be responsible for fixing the issues you detected, then you need to provide your audience with a complete roadmap for improving their content, in other words. If you are going to be working to improve the content yourself, you only need to summarize what this roadmap will entail. However, it's never wrong to be more thorough than you need to be. After all, even if you are going to improve the content yourself, a full roadmap can serve as a checklist for your work after the initial audit.

Wrapping Up: Content Auditing Is a Complicated Activity!

As we were drafting this chapter of the book, we quickly realized it would be the longest and most complex chapter. And that it needed to be. Content auditing is one of the most complex activities content strategists engage in, and it is a core activity for so many other processes. A sound content audit can provide necessary data for improving an organization's content. And, conversely, organizations that don't regularly conduct content audits seem destined to end up with unruly, ineffective content that will only get worse over time.

And, as we mentioned earlier in this chapter, we have never done a content audit for an organization who could accurately tell us the amount of content associated with their channels. Unsurprisingly, none of these organizations had ever audited their content in a systematic fashion. Though we don't have data to support this claim, from talking to dozens of other content strategy practitioners over the years, we believe that the *majority of organizations* are not conducting regular content audits.

If we are correct in our estimation of the size of this problem, then it means that there are literally thousands of organizations out there in desperate need of a content audit. Most organizations tend to produce content in a piecemeal fashion by responding to specific exigencies. They write documentation when a new product or service is launched. They draft blog posts when their rate of website visitors is falling. They create newsletters when they're trying to retain existing customers. And so on. But no one is *assessing* all of this content after it's produced. And thus, no one in the organization really knows what the content is doing for the organization (or failing to do) after it's initially produced.

And if this process repeats itself over years or decades, then the organization's content will only grow and become more and more unruly and undisciplined as time goes on. And in our personal experience, this process tends to continue, unabated, until the ineffectiveness of content itself becomes a primary exigence. We've dealt with clients who had lost 50% or more of their website audience by the time they came to us. We've dealt with clients who refused to believe that their website had ballooned to over 800 pages in the course of a decade, despite the data we showed them.

Imagine all the time, effort, and other resources lost to ineffective content! And thanks to widely available, relatively inexpensive tools like consumer-grade content management systems (i.e., WordPress, Drupal, Joomla!), the problem seems to only be getting worse over time.

But there is hope. There is a reason why content strategy is such an in-demand profession right now. As new channels for content are launched each year and as these channels get loaded up with content, more and more content-focused professionals will be needed to tame all of this information. The humble content audit can save an organization from crashing and burning due to content-related issues. We hope this chapter has been useful for introducing this core method.

Getting Started Guide: Doing Your First Content Audit

If you've read through this chapter, you may be feeling intimidated by the complexity of content auditing, but take heart! We've developed this discussion guide to walk you through your first audit. It's important to remember that like any professional process, content auditing is something that needs to be learned overtime. You won't be an expert at it the first time you do a content audit, or the tenth time. In fact, over years of auditing, developing, and publishing content, we've found that each content audit comes with its own unique challenges.

There are certainly repeatable steps that can be used in almost any audit, however. So, we're presenting these steps here to you to help you get started!

And remember: the most important thing about content auditing is that it is goal driven. It is much better to do a limited audit that is driven by a very specific goal than to attempt to audit a large amount of content without a clear goal in mind. You can bet that professional content strategists who do content audits of large amounts of content have incredibly specific goals in mind.

That's why the first step we recommend is setting goals for your audit. Follow these steps to start your first content audit:

1. Before starting your audit, discuss initial goals with your stakeholders and agree upon specific goals that meet the MAST criteria discussed in this chapter. We can't emphasize enough how

important this step is. Content audits can easily spiral out of control if they aren't driven by measurable, achievable, simple, task-oriented goals. Also discuss with stakeholders the total amount of content that will be audited. You should be *very* concerned if a client asks for a comprehensive audit of all their content. Depending on the size of the organization, this could include the equivalent of thousands of pages of content! Again, we highly recommend starting with the low-hanging fruit content, the content that the organization is most concerned about, the content that is currently impacting organizational goals the most.

2. Download the Example Content Audit Spreadsheet. The easiest way to do this is to go to the link (*https://bit.ly/30jUQaT*), click File, and then click Download. You can download the template as an Excel document, an OpenDocument, or a CSV (comma-separated values). If you're signed into your Google account, you can also go to File > Make a Copy and create a Google Sheets version that you can edit directly that way.

3. Use the guidelines in this chapter as well as the tips in the template to conduct your audit. Make sure you complete the following tasks as you do so:

   a. Conduct a content inventory
   b. Develop a rubric for assessing content
   c. Assess content
   d. Create findings from your audit
   e. Write a content audit report

2. We recommend rereading the sections of this chapter that detail each of the aforementioned tasks as you do them for the first time. You might also want to take notes as you reread in order to make sure you're clear on how to conduct each step.

3. Once you've completed all the steps of a content audit, you're ready to present your findings to your client and discuss how to improve their content! A natural next step after conducting an audit is to develop content models for the content you've uncovered, which is discussed in Chapter 6, and then to assemble a full a content strategy plan, which is discussed in Chapter 7.

References

Flesch reading ease and the Flesch Kincaid grade level. Readable. (2021, July 9). Retrieved October 15, 2021, from https://readable.com/readability/flesch-reading-ease-flesch-kincaid-grade-level/

Gove, J. (2019, February 12). *What makes a good mobile site?* Google. Retrieved October 1, 2021, from https://developers.google.com/web/fundamentals/design-and-ux/principles

Johnson, J. (2021, September 10). *Internet users in the world 2021.* Statista. Retrieved October 1, 2021, from www.statista.com/statistics/617136/digital-population-worldwide/

Johnson, J. (2021, October 08). *Search engine market share worldwide.* Statista. Retrieved October 15, 2021, from www.statista.com/statistics/216573/worldwide-market-share-of-search-engines/

Siang, T. Y. (n.d.). *What is interaction design?* The Interaction Design Foundation. Retrieved November 11, 2021, from www.interaction-design.org/literature/article/what-is-interaction-design

Smart goals: Definition and examples. Indeed Career Guide. (2021, September 21). Retrieved September 30, 2021, from www.indeed.com/career-advice/career-development/smart-goals

The beginner's guide to SEO: Search engine optimization. Moz. (n.d.). Retrieved October 15, 2021, from https://moz.com/beginners-guide-to-seo

Wroblewski, L. (2011, October). *Mobile first.* LukeW. Retrieved October 1, 2021, from www.lukew.com/resources/mobile_first.asp

Further Reading

Frick, T. (2021, June 22). *Content audit: A step-by-step guide.* Mightybytes. Retrieved December 9, 2021, from www.mightybytes.com/blog/how-to-run-a-content-audit/

Kaley, A. (2020, September 27). *Content inventory and auditing 101.* Nielsen Norman Group. Retrieved December 9, 2021, from www.nngroup.com/articles/content-audits/

Land, P. L. (2014). *Content audits and inventories: A handbook.* XML Press.

McCormick, K. (2021, August 10). *How to perform a content audit: Full guide +6 free templates.* WordStream. Retrieved December 9, 2021, from www.wordstream.com/blog/ws/2021/08/10/content-audit

Patel, N. (2021, August 28). *How to run a content audit.* Neil Patel. Retrieved December 9, 2021, from https://neilpatel.com/blog/content-audit/

Petrova, A. (2020, November 25). *The step-by-step guide to conducting a content audit in 2021.* Semrush Blog. Retrieved December 9, 2021, from www.semrush.com/blog/content-audit-for-content-marketing-strategy/.

Spencer, D. (2014, October 16). *How to conduct a content audit.* UX Mastery. Retrieved December 9, 2021, from https://uxmastery.com/how-to-conduct-a-content-audit/.

6   Content Modeling

Now that we have a basic understanding of several of the processes central to content strategy, including audience analysis, identifying content types and channels, and content auditing, we can work on templating out what should go into specific content genres, a process known as *content modeling*.

What Is Content Modeling?

Before we can begin creating content models as a content strategist—an important part of creating content that is both effective and efficient, we have to understand what we mean by *content modeling*. As mentioned in Chapter 2, we can think about content modeling in its most basic sense as *the process of creating structures and frameworks that content must adhere to*. In other words, content modeling allows us to create an outline as to what our content should look like, how it should be formatted, and how it should be written. This allows all of our content of a specific *genre*, or type, to look the same, maintaining a level of professionalism and consistency across our organization and its identity.

However, it should be noted that while a content model may seem simple in theory, it can start to become a bit more complex as you attempt to manage more and more genres of content.

So, we've used several terms to talk about classifying content: content type, content channel, and content genre. To break this down a bit for you:

- Content type: the basic components of content, such as image, link, text, and button
- Content channel: the way content is delivered to an audience, such as website, social media feed, email, and book
- Content genre: the specific type of content that connects content type and content channel, such as *an Instagram post advertising a fundraiser for a non-profit*, or *the homepage of a hospital's website*

DOI: 10.4324/9781003164807-6

These terms go from most basic to most complex, essentially. Content types are the basic units of content. Content channels are the delivery mechanisms. And content genres are the specific names for the ways content types are delivered to an audience (so, content types + content channel).

We've tried to use terms consistently throughout, but the truth is that, as an emerging field, many of the terms in content strategy are in flux. And sometimes practitioners, or even academics, use terms inconsistently. When in doubt: communication is what matters. Use terms that make sense to you, your audience, and your fellow content creators.

Why Matching Content Types to Channels Is Important

Since we can think about content modeling as developing streamlined outlines for creating different kinds of content, it's important to understand the different genres that we may be working with as a content strategist. Each genre of content serves a very specific purpose, as individual content types are often distributed across several different channels. At the same time, content types must maintain some level of consistency across them. Your job as a content strategist is to make sure that each genre (and their corresponding channels) has a content model in place to make the process of developing content as efficient as possible. Doing so means that you'll be able to simply create content and publish it, rather than getting bogged down formatting, designing, and crafting content from scratch each time.

One way to think about this is branding. How do you know when you're seeing content online from your favorite company, non-profit, or university? Besides there potentially being a logo attached to the content, you probably recognize the style of the content as belonging to that organization. Branding goes far beyond including a logo in a webpage or Facebook post. It includes the way the organization uses all of its content, from updates about the latest services being offered to event reminders.

Content is, in many ways, the lifeblood of organizations. It allows them to connect with external audiences in a way that clearly communicates that the organization is a coherent whole.

Now, you've probably seen the opposite happen as well: when content is *not* consistent across channels it can be confusing as an audience member. If you're a college student, you may have gotten an email from your university that confused you because it didn't indicate that it was from an authoritative sender, such as a professor, dean, or provost. If you're working in the corporate setting, you may have seen a blog post on the company website written by another department that conflicted with information you had recently seen in your own department.

These are the *content silos* we talked about earlier (in Chapter 1): when content creators writing within organizations don't have a shared plan, a shared vision of what content should be, then efforts are going to get duplicated, informational wires are going to get crossed, and content is going to lose consistency across channels.

Content modeling helps avoid this by communicating to everyone within an organization what each genre of content used by the organization should contain. People in marketing don't have to guess about what a landing page on the company website geared toward prospective customers should contain, because there's a template available for landing pages. University professors don't have to guess how they should communicate to college students who are misbehaving in class because there are guidelines out there for what such emails should contain.

Now, in our experience, this kind of careful structuring of content across an organization simply doesn't happen very often. What's far more common is the siloing of content we discussed in Chapter 1. Marketers within an organization are creating landing pages from scratch because they don't know there's a landing page template the company uses. Then they find out when they're ready to launch the dozens of landing pages they've created and all their work is wasted, as well as the company time and resources that they've used. Any university professors reading this book are laughing at the example of a guide for communicating with misbehaving students, because typically universities don't create simple guides like that, relying instead on very complicated policies that are difficult to understand when you're doing something like composing an email to a student. This is why many university professors who are communication experts end up creating their own templates for things like student emails, grade reports, and feedback on project drafts.

It's also important to understand that each channel that an organization is publishing content through (social media, email, blog, etc.) may have a different audience, or, more likely, multiple audiences. You may have prospective donors reading your non-profit blog posts about your fundraising efforts, but volunteers may also consume that content to learn how their efforts are making an impact on your organization. Content is also often used for a very specific purpose by each audience persona. A first-time donor to a non-profit is not going to use the organization's website the same way that a repeat donor is.

Your content model needs to represent all this complexity. It needs to explain what the purpose of each genre of content is, who it's for, and what it does for the organization.

In this chapter, we identify some of the most popular content genres that you may end up modeling for an organization, describe what a content model looks like, and describe how you can streamline the process for adapting content channels.

Common Content Strategy Genres

As we have mentioned throughout this book, while we try to represent some of the most popular situations, content channels, content types, and content genres that organizations use, we understand that there will be exceptions for your specific organization or industry. Our goal here is to provide examples of content genres that are popular in the world of content strategy, as well as provide you with the tools to be able to identify and craft your own content models in the future.

We cover this resource more thoroughly in the next chapter, but to aid you with the content strategy process, we've created a Content Strategy Plan template to help you: *https://bit.ly/3iBLqgy*. Part of that template is a list of content models. There are a few more than what we cover in this chapter, but the main value-add of the template is that it's fillable. So, if you're starting to plan your content models in this chapter, you might want to download the template as a Word document, an OpenDocument, or an RTF. If you're signed into your Google account, you can also go to File > Make a Copy and create a Google Doc version that you can edit directly that way.

Again, however: we cover the ins and outs of using the template in Chapter 7, which talks about the content strategy planning process in-depth.

Blog Post or Webpage

The blog post and webpage are two of the most common content genres that organizations create. And these two genres can quickly grow out of control if not managed efficiently, effectively, and *consistently*. We've consulted with many an organization who vastly underestimated the amount of content on their website. This is partially the fault of modern-day content management systems (CMSs) like WordPress, Joomla!, and Drupal that enable easy publishing of content, but contain no default mechanisms for *auditing, analyzing, or maintaining content once it's published*. We've had clients look in shock at a web crawler report on their website showing hundreds of pages they didn't even know they had.

To avoid issues like this, you want to make sure that no matter who in your organization is writing a blog post or creating content for a webpage, that all of this content has the same look and feel for your audience.

You might be rightfully asking as this point: "but aren't blog posts and webpages very different genres?" In some ways they are, yes, but in many ways they're very similar. The way modern websites are structured means that any piece of content, whether it's a blog post, an image, or a webpage, is attached to the root domain of the website by a link, such as http://yourdomain.com/this-is-a-blog-post. So, from a technological standpoint, all website content is structured roughly the same way.

The main difference between a webpage and a blog post is not the way it's structured, then, but the purpose it serves for an audience. A blog post is meant to be more time-sensitive than a webpage. An About Us page on a website is most likely going to remain in its rough form for the duration of the website's existence, for example. New personnel might be added to the page, but the site will most likely always have a page that tells website visitors about the organization, what it does, and who works for it.

Blog posts, on the other hand, are published on a semi-regular basis. They range from limited-time content, such as an event announcement, to evergreen content, such as a list of ways to make use of a product or service. That's really the only difference, however. The . layout of blog posts and webpages have even begun to merge together a bit, with many of them following the "landing page" idea that every page on your website should immediately orient visitors to the purpose of your website and inspire them to take action.

So, what would a content model look like for a blog post or webpage? We have to start by asking ourselves: what are the different parts of a blog post or webpage? Or, in other words, what content types should our content model contain?

Let's start by thinking about some of the most important parts of a blog post or webpage. First, we probably want to make sure that we have a very relevant title. By relevant we mean one that is not only eye-catching for our audience, but that also includes a keyword that is essential to the topic addressed. Remember, as we discussed in Chapter 5, search engines are key channels for any form of web content, and search engines are driven by keywords.

Next, we need to think about an introductory paragraph for the post or page. This intro should once again utilize a keyword that is central to the topic and help explain what the overall purpose of the post or page is.

After the intro, we move into the main body of our post or page—where the bulk of content is going to be housed. Additionally, it's important to note that most SEO experts claim that the most effective blog posts are somewhere between 300 and 1,000 words, though longer-form content of up to 2,500 words has been shown to outrank shorter posts (Bunting, 2021). As webpages are structurally the same as far as search engines are concerned, these lengths should hold true for them as well.

The main body of the post or page should also focus on the keyword we've identified. The post or page should also contain at least one image. Search engines like rich media in their website content. In that way, they're similar to many human users.

Lastly, we wrap up our post or page with a conclusion. Just like your typical essay, we want our conclusion to be brief and to recap the overall topic for the reader. Like all elements of the post or page, it should focus on our keyword. We should also include a call-to-action that involves your organization or prompts the reader to take further action, such as interacting with additional content, contacting the organization, or subscribing to future updates.

In other words, our content model for a blog post or webpage would look something like Figure 6.1.

What you see here is thus a basic outline, a blueprint if you will, for everyone to follow within an organization that lays out the *content types and how they should be assembled when creating webpages or blog posts*. The more specific you can get with your content model, the better, however. See the following for an actual example from a past client for a specific blog content model focused on women's health:

- Title

  - o Use keywords related to women's health, including "women's health issues," "menstruation problems," "lower back pain during pregnancy," [. . .].
  - o Make titles explanatory and educational, that is, "How to Deal with Lower Back Pain During Pregnancy" [LINK].

- Intro

  - o Include a photo at the beginning of the post from the following Shutterstock account [CREDENTIALS].
  - o Include an alt tag for the image that includes the keyword.
  - o Use the same keyword you used in the title in your first sentence. Start the post by introducing the topic in non-medical terms. Explain what you will cover in the post and why the topic is central to women's health issues.

- Body

  - o Use at least one link to another website from the following list: [LINKS].
  - o Explain symptoms related to the issue, possible treatments, and a list of resources where the reader can get more information. Also include at least one prevention tip gleaned from information from one of these sources: [LINKS].

- Conclusion

  o End with calls to action that refer the reader to our newsletter sign-up [LINK] and invite them to schedule an appointment [LINK].

- Once a patient information blog post is written, schedule it in Word-Press [CREDENTIALS] for the following Wednesday at 9 a.m.
- When you do so, check the Yoast SEO plugin readout, fix any SEO errors, and save the changes to the post.

As you can see from the previous example, this fully-fleshed out content model includes more specific information than our basic example in Figure 6.1. The difference is that Figure 6.1 is simply a starting point whereas the content model for a blog post on women's health is *a content model that an actual organization used*. That's why it contains information specific to the organization's goals and audiences. You'll see a reference to specific keywords as well as a reference to a Shutterstock account, for instance. You'll see a reference to scheduling the post once it's written.

o Title
  - Use a keyword that will be central to the topic you're writing about. If you're not sure what good keywords are for your particular organization, I highly recommend using SEO Book's Free Keyword Tool (http://tools.seobook.com/keyword-tools/seobook/).
o Intro
  - Use the same keyword you used in the title in your first sentence.
  - Explain what the overall purpose of the post is.
o Body
  - Use at least one link to another website.
  - Use at least one image. I like to use Flickr's Creative Commons (http://www.flickr.com/creativecommons) or Pixabay (http://pixabay.com/) for this. If you use Flickr, be sure to select "available for commercial use" from the search options.
  - Include an alt tag for the image that includes the keyword.
  - Explain a process, interesting fact, or personal story that highlights your expertise as a [TYPE OF ORGANIZATION].
  - Make sure your overall word count is at least 300 words and no more than 1000 words.
o Conclusion
  - Include a call to action that involves your organization.

Figure 6.1 Example blog post or website content model from the content strategy plan template

Some common types of blog and webpage content models you might want to create include:

- Website landing pages for a specific campaign, product, service, or location
- Blog posts that seek to be educational, information, entertaining, or persuasive
- Product pages, such as those that go into an online storefront (see Chapter 4 for an example)
- About us pages that explain an organization's mission, philosophy, and show the real, live people involved
- Donor, customer, or volunteer spotlight posts or pages

Essentially, a fully operational content model should tell anyone in the organization exactly what they need to do when creating that type of content. There will always be some room for interpretation, such as what specific keyword to use, what images to use, what specific language to use throughout the post. But the specific content types that the genre needs to contain, how those content types should be arranged, and the overall purpose of the genre? That needs to be decided beforehand.

Article in a Structured Database

Closely related to the blog post or webpage content model is the model for an article. While a blog post may be something that is meant to be a timely piece of content added to a website, an article is often broader in classification. Articles can appear in lots of different channels, for instance, such as newspapers, magazines, customer help forums, and many others. In fact, articles have so many different purposes and audiences, it can be difficult to classify different articles as the same genre. A news article in a magazine like *Newsweek* is going to be very different than an article in an academic journal such as *Technical Communication*, the peer-reviewed journal of the Society for Technical Communication. An article in an interdepartmental newsletter about an upcoming company picnic is going to be very different from an article written for an online magazine on tractors.

Rather than try to define a content model that would apply to all articles, then, we want to use this broad genre to discuss another facet of content modeling: *structured authoring*. According to the website of Oxygen, one of the most popular structured authoring tools for technical communicators, structured authoring is:

A standardised [*sic*] methodological approach to the creation of content incorporating information types, systematic use of metadata, XML-based semantic mark-up, modular, topic-based information architecture, a constrained writing environment with software-enforced rules, content reuse, and the separation of content and form. (*Structured authoring*)

What does all that mean? Let's break it down.

First, structured authoring is a standardized methodological approach to content modeling, meaning that it's a standard best practice for people who deal with technical content, such as technical writers, technical editors, and content strategists in technical fields like engineering, medicine, or IT. Next, structured authoring incorporates different types of information, meaning that it includes the storage of a variety of content types—for example, link, image, body text, title. It also incorporates the use of metadata, or "structured reference data that helps to sort and identify attributes of the information it describes" (Kranz, 2021). You've seen metadata, whether you realize it or not, every time you use a search engine. The titles, descriptions, and links to websites that come up in search engine results pages? Yep's that's all metadata. It's reference data that tells you what's on each page, so you don't have to scroll through all of the content of each page.

Next, structured authoring uses *XML*-based semantic mark-up. XML, short for Extensible Markup Language, is a markup language, meaning it's similar to HTML. However, whereas HTML comes with predefined tags (such as <header>), XML allows you to make up your own tags (*XML Introduction*). This means you can use it to create whatever kind of classification scheme you want to.

Finally, there's a reference in the definition to a "modular, topic-based information architecture," a "constrained writing environment with software-enforced rules," "content reuse," and "the separation of content and form." What this means, essentially, is that articles that are stored in a structured database are organized in a way that you can't change how they're structured without changing the whole system. And they use software to keep them this way.

That's a lot to digest. In fact, there are entire articles, websites, and books dedicated to structured authoring (several of which we cite in the Further Reading section) that can give you a deeper look at all the interesting things you can do with structured authoring. But for the sake of *this* book, let's summarize all the information we just parsed and give you the abridged version. In short, structured authoring is the creation of content using XML-based markup to help with content reuse.

Think of it this way: the first content model we showed you (for blog posts or webpages) was very simple. It was largely a list of content types that can be stored anywhere, in a Word doc, in a PDF, in a Google doc, in the pages of this book, and so on. And that's great if you only need one content model for your entire website.

But what if you need a content model for 50 different websites? Or a model for a website, mobile app, enterprise software package, and company intranet? What if you need to store a whole lot of information, publish it on demand, and then export it to a bunch of different content genres? And by a "whole lot," we mean tens of thousands of pages worth. What then?

This is where structured authoring comes in. It's largely a process for dealing with issues of scale. Creating a content strategy plan for a single website is relatively easy. But creating a plan for an entire company that has 10,000+ employees and thousands of different content genres? That's really hard. You need technology to manage all that, because there's no one human being who can do all that, or even one team of human beings.

So, a tool like Oxygen (and we'll get more into tools in Chapter 13) enables you to store all that information in a *structured* format, which actually means *stripped of all the styles associated with a specific genre*. You store your information in a very basic format like the following (which is an example of common XML tags):

Title: the title of your article
Description: a brief description of your article
URL: the URL of the article
Author: who wrote this article
Audience: who is the target audience for this article
ArticleBody: the entire article's body text

Then, when you want to see all the body texts of articles written by so-and-so, you can find that information, even if it's exactly 439. If you want to see all the content, in any form, that was written for the audience of *internal stakeholders* or *repeat customers who purchased a subscription in the past 12 months*, you can find it. And not only can you find it, if you've set your structured authoring tool up right, you can publish it to any of the channels your organization uses *without losing the formatting*.

If you've ever tried to get content out of a PDF or to convert an image saved in a Word doc to a standalone file, you know that content doesn't typically change forms very easily. That's the power of structured authoring. Pretty cool, right?

Not all content strategists deal with structured authoring, but many in technical fields do. Like many of the topics introduced in this book, it's something you ought to know about if you're interested in this field, but you might not make use of on a daily basis. It all just depends on where your career path takes you.

Social Media Post

While there are many different social media platforms that organizations can make use of, many use a similar structure when sharing content across accounts. Remember that a content model should be simple enough to work well in a variety of situations, but specific enough to maintain consistency across situations.

As far as purpose, though social media are useful for a variety of reasons, one important feature for organizations is their ability to point audience members to a website. Organizations use social media to encourage people to learn more information, buy a product, subscribe to a service, and so on. Inspiring action, typically called *engagement* in social media parlance, is key. We've worked with a variety of organizations over the years who've used social media to send endless, dry updates, such as event reminders, product releases, or requests for donations.

At the same time, each social media post's content model should use the same basic message structure:

- An interesting fact, quote, or call-to-action related to _____ [topic]
- A link to a source [your organization's website or an influencer's website]
- An image

Again, this is a basic content model to get you started. With any social media campaign, you'll want to template out the different kinds of posts you'll be sharing as well as the different components that are specific to a given channel. Instagram and Twitter use hashtags, for instance, while LinkedIn doesn't. You can use hashtags on Facebook, but they're more for flavor than searchability.

Some common types of social media content models you might want to create include:

- Channel-specific models (e.g., Facebook, Twitter, Instagram, LinkedIn, YouTube)
- Campaign-specific models (e.g., selling a product or service, advertising an event, requesting donations)
- Audience-specific models (e.g., donors, volunteers, customers, internal stakeholders, potential employees, potential employers)

And don't do like many of our clients and make every post about selling or advertising. Include some content that is purely informational and/or entertaining. In fact, the vast majority, around 80%, of your social media content should be purely for engagement. And only 20% should be about selling or advertising. In digital marketing parlance, this is referred to as the *80/20 rule* (Sullivan, 2021). Remember that modern audiences are very sensitive to pushy sales language and will typically disengage if they feel overwhelmed by it.

Emails (for External Audiences)

If your organization is one that sends out a lot of emails to consumers, then it's also important to think about a content model for email. Tools like MailChimp, which is our recommended tool for email content (https://mailchimp.com/), often come bundled with some great email templates (https://mailchimp.com/features/email-templates/). As usual, though, you should develop a content model that clearly explains the purpose of emails your organization sends out to maintain consistency.

A lot of the *content* that goes into this model may look similar to that of a blog post or webpage. This is because modern email programs like Outlook and Gmail format email content in ways that are similar to how web browsers format website content. However, the key difference here is that we are delivering content straight to our audience's inbox—they don't have to come to our website to read it. That means the subject line becomes key.

In our content model for an email, we can think of the subject line like the title of a blog post or webpage. It needs to be something that pertains to the overall topic of the email but *needs to grab the attention of our audience* as they scroll through their inbox.

Next, we're going to want to make sure that we include at least 1–2 images, but no more than 2.

After that, we want our content to contain information that our audience can't find anywhere else—something that is exclusive to those that are subscribed to our mailing list.

Lastly, just like our blog post or webpage, we want to end our email with some kind of call-to-action that will bring them back to our website. This CTA should prompt them to take further action such as reading more about a specific topic, contacting our organization, or staying engaged on another channel, such as social media.

Some common types of email content models you might want to create include:

- A VIP newsletter for regular customers or donors
- A basic outreach email for people who have just engaged with your organization for the first time
- An email for people who you haven't heard from in a while
- An email for people who just made their first purchase, donation, and so on
- A thank you email for volunteers or other types of supporters

Other Common Content Models

The content models we've presented in this chapter are just a handful of the most common ones used by content strategists. As we've tried to stress, however: the more specific you get with content models, the better. They should be specific to a purpose, an audience, and an organization. Generalized models like the ones we've presented here will only get you so far.

At the end of the day, a content model should help you create, audit, and improve content. It should serve as a framework or outline for working with specific content genres. If the model is too specific, you risk losing consistency across channels. If the model is too general, you'll spend too much time reinventing the wheel every time you develop content.

Content modeling is more an art than a science. It's a planning activity that helps you think about content on a larger scale than when you sit down to just create a single piece of content. And it's an answer to the question of how modern-day content strategists can manage all the different content genres out there.

That being said, here are some other common content models we've seen used by content strategists:

- Ebook
- Internal documentation
- RFP (request for proposals)
- Help documentation
- User instructions
- Product specification
- Employee handbook
- Infographics
- Case studies
- Whitepapers
- Technical reports
- Podcasts
- Video tutorials
- Webinars
- Online courses

In truth, there are simply too many content genres in use right now to include a complete list. That's why in the final section of this chapter, we provide tips for doing what you'll end up doing a lot of the time as a content strategist if your experience is anything like ours has been: creating your own content models.

Getting Started Guide: Creating Your Own Content Models

Although the genres that we have mentioned here are some of the most popular content genres you may work with, we also understand two things: (1) that the way that your organization publishes a specific content genre may differ strongly from what we've described in this chapter, and (2) that your organization may use content genres that we haven't even mentioned. Add to this that new content genres are being born all the time (don't even get us started on TikTok!) and you'll need a process for creating your own content models.

That is why we identify some key heuristics in this section to allow you to create content models of your very own from scratch.

Don't be scared about the word "heuristics" here. By using that word, we are mostly focusing on specific guidelines or "rules of thumb" on how to get started. Here are the steps to get you started with content modeling.

Step 1: Identify the organization that you want to create a content model for

This is probably the easy step—it's probably the organization you work for! But, if you're a content strategist that works with multiple clients, it's always helpful to remind yourself about the organization, what their goals are, and what their challenges are. If you're learning about content strategy in school or another educational venue, you may need to locate a local organization, such as a non-profit, to practice content modeling.

Step 2: Determine the content genre that you need a content model for

Next, you need to identify, in as much detail as possible: what is the specific content genre you're trying to model? Some questions that might help you include:

- What is this content being used for?
- What content types are involved?
- What are the goals of your audience for the content?
- What is the content going to do for your audience?

Step 3: Determine the content channel that content will be distributed to

You need to understand where this content is going to be published. Is this content strictly going to be used as a blog post? Is it going to be used in social media posts? If so, which social media channel? Is it going to be used across multiple channels?

These questions will affect the content types you need to include (or exclude) from your content model, so you need to determine this before you get started building your actual model.

Step 4: Research similar content genres from competitors

As we explained in Chapter 4, one of the most common forms of research content strategists engage in is competitive analysis: they locate other organizations who are using content in an innovative way and imitate them. It might be the case that you've already created content like the type you're modeling in the past, but never had a content model for it. Or you might be creating a genre you've never created before.

Regardless, competitive analysis can fuel your understanding of a particular content genre like almost no other research method. Seeing the different themes, formats, and content types that other organizations use can help you get a firm grasp on a particular content genre. And don't be afraid to look up best practice articles by thought leaders regarding a particular content genre.

Step 5: Begin making your content model

Next, using your knowledge of previous content you've made for your organization, the research from your competitive analysis, and your understanding of the purpose of your content, your audience, and their goals, it's time to try making your content model.

There are basically two ways to do this:

- As a plan that exists in its own document, such as a Word doc, PDF, or Google Doc
- As a feature within software, such an authoring tool, marketing tool, or website CMS

We typically recommend using the first method almost all of the time. Tools change. If you put all of your content models into software and then your organization decides to discontinue the software, then you'll lose all the content models you created within it. At the same time, we encourage people to use tools for content modeling, because that's how a lot of content is stored and published. If all your organization's website content is stored in WordPress or Drupal, you need to account for that in your model! We talk more about the role of tools in content strategy in Chapter 13.

Step 6: Testing your content (and revising your model)

As we discuss in Chapter 10, our chapter on ensuring usability and accessibility, after you've created your content model and used it to create some content it's important to *test that content with an actual audience*. You may find that some things are working well while others aren't, and you need to make a change as to how you craft future content. If that's the case, you may need to revise your content model to ensure that all content moving forward is as effective as it can be, and that everyone using that content model knows what to avoid.

References

Bunting, J. (2021, January 20). *How long should your blog post be? A Writer's Guide.* The Write Practice. Retrieved March 22, 2022, from https://thewritepractice.com/blog-post-length/

Kranz, G. (2021, July 12). *What is metadata and how does it work?* WhatIs.com. Retrieved March 22, 2022, from https://whatis.techtarget.com/definition/metadata

Structured authoring. Oxygen XML Editor. (n.d.). Retrieved March 22, 2022, from www.oxygenxml.com/dita/styleguide/webhelp-feedback/Artefact/Authoring_Concepts/c_What_is_Structured_Authoring.html

Sullivan, M. (2021, January 21). *The 80/20 rule, explained.* BrandMuscle. Retrieved March 23, 2022, from www.brandmuscle.com/resources/the-80-20-rule-explained/

XML Introduction. MDN Web Docs. (n.d.). Retrieved March 22, 2022, from https://developer.mozilla.org/en-US/docs/Web/XML/XML_introduction

Further Reading

Gibbon, C. (n.d.). *Content models.* Cleve Gibbon. Retrieved March 22, 2022, from www.clevegibbon.com/content-modeling/content-models/

Ismail, K. (2018, January 3). *Content modeling: What it is and how to get started.* CMSWire. com. Retrieved March 22, 2022, from www.cmswire.com/content-strategy/content-modeling-what-it-is-and-how-to-get-started/

Lovinger, R. (2012, April 24). *Content modelling: A master skill.* A List Apart. Retrieved March 22, 2022, from https://alistapart.com/article/content-modelling-a-master-skill/

O'Keefe, S. (2021, December 17). *Structured authoring and XML.* Scriptorium. Retrieved March 22, 2022, from www.scriptorium.com/2017/04/structured-authoring-and-xml/

Pope, L. (2021, November 15). *Content model: Why you need one and how to make your colleagues take notice.* GatherContent. Retrieved March 22, 2022, from https://gathercontent.com/blog/why-you-need-a-content-model-how-to-make-colleagues-take-notice

Rose, R. (2020, September 2). *Think strategically about your content model.* Content Marketing Institute. Retrieved March 22, 2022, from https://contentmarketinginstitute.com/2020/09/think-strategic-content-models/

Swisher, V. (2016, June 6). *Structured authoring: Without it, you're spending way too much time creating content.* Content Rules, Inc. Retrieved March 22, 2022, from https://contentrules.com/structured-authoring-save-time-creating-content/

7   Assembling a Content Strategy Plan

What Is a Content Strategy Plan?

A content strategy plan is a plan that includes a written strategy for all of the activities that an organization will engage in to carry out their content strategy. It is typically written up as a formal document that is shared within an organization so that everyone creating or managing content within the organization is on the same page. Sometimes it exists as a kind of manual that is published in PDF format and stored somewhere everyone can easily find it. Sometimes it is embedded within technology, like a content management system. Sometimes it's a little of both.

From previous chapters, you probably already recognize the utility of drawing up a formal content strategy plan. Content strategy involves a lot of different components and activities, and it can be easy to let things slip by the wayside if you're not careful. Sure, you audited the organization's website last year, but what's changed since then? Are you keeping up with the organization's blog? Have you checked the content management system for updates to articles in the user help forum that require a response?

Content strategists are jacks or jills of all trades and so they need a plan that will help them keep on top of all the little things they do for an organization. Managing content can also feel a lot like herding cats. If content creators aren't apprised of the standards for content they're creating on a daily basis, then this creates more work for the content strategist, forcing them into the roles of editor, reviser, and proofreader. Even with standards, it can be challenging to get everyone in the organization to follow them.

The other reason to create a plan has more to do with organizational sustainability. Like any professional, content strategists can move on from a given organization. If they don't create a detailed plan of what they're doing, then the next person to fill their position within the organization will essentially be starting from scratch. This can create a very rough transition for someone unfamiliar with all the inner workings of an organization's content flows. What channels does the organization use? What technologies? What

DOI: 10.4324/9781003164807-7

content goals does the organization have? Who are its core audiences? What is the content strategist's scope of responsibility?

Without answers to questions like these articulated within an easy-to-follow planning document, a content strategist may spend weeks or even months just getting up to speed.

It also bears mentioning that because content strategy is still a very nascent profession, much less common than roles like technical writer, IT manager, or web developer, many organizations operate without a content strategy plan. The authors of this book have consulted with organizations ranging from tiny non-profits to large corporations over the years and have yet to encounter a single client organization with a content strategy plan already in place.

Not having a content strategy plan leaves content goals under the purview of individual content creators. This often leads to what Ann Rockley & Charles Cooper (2012) have called "content silos" (p. 133), or barriers between individuals and departments within a single organization. Marketers create content to promote the organization's product or service. Managers write reports about sales and client retention. Sales professionals create scripts for talking to customers. Customer service and support people create and manage content in support forums.

But who is making sure all this content is unified, cohesive, and coherent? Who is tracking its alignment with organizational goals?

If the organization doesn't have a content strategy plan, then the answer is probably: no one. A well-established content strategist once told one of the authors of this book that they had consulted with a major company in the technology sector, the kind of company with tens of thousands of employees. They were asked to do a content audit of the company's website to identify strengths, weaknesses, and opportunities for improvement within the content of the site. This website contained thousands of pages of content and hundreds of thousands of words within those pages.

After an exhaustive audit that took over three months to complete, the strategist had to report to the company that only about 10% of their website content was actually accomplishing anything at all. All of the rest was "dead content," meaning content that was hopelessly outdated, not archived by search engines, and was rarely or never visited by audience members.

When the company's CEO asked the strategist how this had happened when the company had such comprehensive company goals and a mission to provide superior customer service, the content strategist pointed to the lack of comprehensive *content goals*. The strategist explained that running a website, or any complex content channel, without goals was like running a company without quarterly reporting. Without consistent goal setting, assessment, and improvement, it was no wonder that the company's website had developed in such a substandard fashion.

Content goals aren't the only essential component of a solid content strategy plan, however. Next we turn to the rest of these components.

The Parts of a Content Strategy Plan

A sound content strategy plan contains the following basic components:

- Goals
- Audiences
- Content channels
- Content models
- Editorial calendars

Each of these components work together within the plan to create a complete picture of an organization's content strategy. In this way, a well-documented content strategy plan is a stand-in of an organization's content strategy. Without such a plan, all of the content-associated activities the organization engages in are based on the assumptions of whoever is in charge of them.

Content strategy plans are thus collaborative documents that need to be owned by the entire organization, not just a section of it. In our example above, it's unlikely that the technology company created thousands of pages of unnecessary content on their website without doing so elsewhere. A systematic audit of the entire organization probably would have revealed redundancy and wastefulness throughout the company, at least when it came to content. A content strategy plan that only a small portion of an organization is paying attention to is no plan at all, in other words, because if only some people who create or manage content follow it, then it's not an organizational plan, but a departmental plan or workflow among select individuals.

This effort to get everyone on the same page throughout an organization typically starts with goal setting, in which specific content goals are articulated across individuals, groups, and departments.

Creating Effective Content Goals

As we describe in Chapter 5, we use the MAST acronym for creating content goals. They should be: Measurable, Achievable, Simple, and Task-oriented. Content goals should be simple and realizable statements of objective facts that describe a specific aim you wish to achieve. It's also best to set goals you know you can reach and then set new ones when you reach or exceed your initial goals.

Next, it's also essential to link content goals to organizational goals. If an organization wants to sell printers to the elderly but is creating social media content that elderly people don't understand or connect to, then this content will not support this goal. Content goals can't solve organizational problems by themselves, but organizational goals often can't be achieved without content goals. The two work together, symbiotically, to move an organization forward.

Imagine all the wasted effort the employees of the technology company cited here engaged in over the years to create thousands of pages of website content without a clear goal in mind! Now imagine what they could have achieved with this amount of effort had they had clearly articulated content goals in mind. At the very least, they could've saved themselves a lot of time and effort, time and effort that could have been redirected to other tasks, like improving customer service or doing more direct outreach to a specific type of customer.

Here's a template for setting content goals to get you started:

- I will perform [TASK] [NUMBER OF TIMES] per [PERIOD OF TIME] until [STATE IS ACHIEVED, e.g., our website bounce rate drops below 60%] with [X AUDIENCE, e.g., first-time website visitors interested in printers].

In order to create goals using this template, the first thing you need to understand is how the organization creates value for its customers or clients.

We've described this to clients as identifying what activities "push the needle." When organizations are struggling to understand what content does for their organization, we ask them questions like the following:

- What makes you unique in the eyes of your customers or clients? Why do they keep coming back to you?
- What metrics do you track to ensure your organization is successful each month or quarter (i.e., sales, donations, repeat customers, numbers of support tickets, website visitors, email opens)?
- How do you know if things aren't going well in your organization? What indicates this to you?
- How often do you meet with managers in your organization to create new goals?

Once you have a sense of what activities push the needle of the organization's goals, you can think about how to tie those metrics to ones you can track with content.

The best thing you can do with content, you see, is build relationships with audiences, but these relationships may help you solve organizational problems. For example, say our technology company we've been discussing had flagging sales last quarter. This is often the time when outside consultants get called in: when there's something wrong. If all the usual metrics are looking positive, why bother changing things?

Of course, this isn't the best approach when it comes to content goals, because content problems tend to build exponentially over time. If website content is being created with no rhyme or reason to it, then this problem is simply going to get worse over time. Before you know it, the majority of your content is doing absolutely nothing but weighing down your website.

Regardless: our experience is that clients start to look at content when they've exhausted every other possible solution. This isn't terribly surprising as most people who run organizations, be they small businesses, non-profits, or large corporations, aren't trained in content strategy. It's easier to focus on the "big" issues like quarterly earnings than it is to think that your blog content is the "real" problem.

In any case, once you have a sense of what the organization's key metrics are, you're ready to identify their Key Performance Indicators, or KPIs. KPIs are the best ways to tell how the organization is doing. They are the "needle" that moves on the gauge of the company's goals. In our running example, flagging sales would be a KPI. Customers aren't buying as many products and that's going to affect the organization's bottom line.

To start off, try to identify content channels, which we discussed in Chapter 4, that are linked to the KPI. If sales are the KPI, for example, then you would probably look at the content channels that are focused on reaching customers.

The next step in setting a content goal that aligns with that KPI, however, is to think through the rest of the plan. So, we're now going to move to audiences. By the time we've led you through the whole planning process, you can circle back and articulate your goals.

Defining Audiences

In order to set sound content goals, you need to know who your content targets. Recall that the best way to conceptualize an audience when doing content strategy is to develop audience personas, which we discussed in Chapter 3. We're assuming at this point that you've done some audience analysis. If not, you'll need to do that before filling out the audience portion of your content strategy plan.

Once you have identified your audience personas, you need to think about which ones to include in your plan. Not all personas will fit with your specific goals. A content strategy plan should be focused, in other words. It should be comprehensive, but also specific.

Say our example technology company has 10 different personas that represent different types of audiences, including customers, investors, employees, and managers. If the most immediate goal is to help with flagging sales, the focus of the initial content strategy plan should be to think about the audiences that are most involved with that problem.

Don't get us wrong, long-term: an organization should absolutely develop a comprehensive content strategy plan. That's typically not how these plans start, however. And it's perfectly okay to start with the most pressing problems first.

So, in our running example, after talking to different managers in the company, perhaps you realize that part of the problem is that new customers are not being reached, particularly younger customers. From your past audience research, you might have realized that the technology company is a known brand among working professionals who are 50+, but younger working professionals simply don't think of the company when they think about their technology needs. This is a common problem among a lot of direct-to-consumer retailers trying to attract younger generations, whose shopping habits tend to differ significantly from their older compatriots.

At the same time, you don't want to alienate the 50+ customer base, because they're by far the biggest consumers of the technology solutions the company offers. So, now you find yourself with a problem very common to content strategy planning: the need to attract a new audience while retaining an existing one.

And a lot of the time, the solution comes down to identifying what content types and content channels can best be used to appeal to specific audiences.

Identifying Content Channels

As we discussed in Chapter 4, *content channels* are means of distributing different *content types* to an audience. If you've forgotten some of the content channels we listed in Chapter 4, feel free to review that chapter before moving forward with your content strategy plan.

The nice thing about content channels is that there are a lot of them and different audiences tend to gravitate toward specific ones. Think about how you personally use social media: unless you're a very unusual social media user, you probably don't spend an equal amount of time across platforms like Facebook, LinkedIn, Twitter, Instagram, Reddit, and TikTok. In fact, if you're like most social media users, you spend the majority of your time on one or two of those channels. Some channels are also larger (i.e., Facebook) than others (i.e., TikTok). Some have broader appeal across audience demographics (i.e., Twitter) and some have more niche audiences that focus on specific ages, professions, or technology habits (i.e., LinkedIn, Instagram, and TikTok).

We like to tell clients that it's far easier to attract a following in a content channel in which a specific audience is already active than it is to attract an audience to a new channel. Some quick research would tell you that it's hard to attract young users to a platform like LinkedIn.

Some channels contain very large, diverse audiences that need to be segmented. Search engines are a perfect example of this. Nearly everyone uses search engines, which are designed to help users find relevant, authoritative information based on specific keywords. The challenge with search engines is not asking "is X audience on there." Unless the audience doesn't have access to a computer, they probably use search engines. The challenge is to identify which keywords they use to find content.

Talking about social media is a great way to learn about *channel loyalty* or the degree to which users tend to gravitate toward specific channels for receiving content. If you are a younger social media user, for example, you may experience distaste at the thought of even visiting LinkedIn, as many of our students do. If you grew up with much more exciting and engaging channels like Instagram or TikTok, LinkedIn may feel hopelessly boring to you.

It's not that there aren't younger users on LinkedIn, it's that if your primary goal is to reach younger users and engage them in a specific way (i.e., via hip, trendy videos), LinkedIn may not be your best method of reaching them. If you want to reach young small business owners, however, then LinkedIn is a great channel for you.

The best way to research channel loyalty isn't to think about our own personal preferences, however, but to look at demographic research. The Pew Research Center tracks the demographics of social media users and adoption in the United States and is a great source to start with (Pew Research Center, 2021). There are lots of social media firms that track this data as well. Doing some research on Google can help you understand audience loyalty when it comes to social media, as well as a host of other channels, such as blogs, webpages, mobile apps, print books, ebooks, and more.

Many times, it's best to do your own primary research into the audience you're creating content for, however. Designing a survey that asks them about the ways they like to receive content is a great way to understand how to reach them.

It's also perfectly appropriate to identify a couple of channels that contain the specific audience you're looking for. Developing a blog about 3D printer hacks that younger people can use for creative projects and then sharing those hacks via examples of projects people have done over Instagram would be an example of hitting two channels at once: search engines and Instagram. As a rule, you should try to make your content do as much work for you as you can, in other words, rather than reinventing the wheel each time you get ready to create content.

As you know from Chapter 6, that kind of planning across content channels and audiences is called content modeling.

Content Models

As we discussed fully in Chapter 6, content models are structures and frameworks that content must adhere to. Content models are outlines that define what our content should look like, how it should be formatted, and how it should be written. This allows all of our content of a specific *genre*, or type, to look the same, maintaining a level of professionalism and consistency across our organization and its identity.

In the context of creating a content strategy plan, content models serve the purpose of identifying the specific genres of content that the strategy pertains to. Remember that the purpose of a content strategy plan is to serve as a guiding document for an entire organization, or at least for a part of an organization (such as a specific department, like technical publication). This means that besides specifying goals, audiences, and channels, you also need to specify the requirements of specific genres in your content strategy plan. Otherwise, it can be easy for individual content creators to create content in a way that doesn't align with the rest of the plan.

Again, for details on how to create specific content models, revisit Chapter 6.

Editorial Calendars

Last but not least, the final component of a solid content strategy plan is an editorial calendar. We like to tell people who are new to content strategy that an editorial planner is simply a fancy name for a to do list. In the context of content planning, an editorial calendar *lays out all the tasks, processes, and deadlines associated with your ongoing content strategy plan.*

There are lots of ways to represent an editorial calendar in a content strategy plan. Sometimes, the calendar exists within a calendar tool, like Google Calendar (*https://calendar.google.com/*). Sometimes it exists within a content management tool such as WordPress or Hootsuite. Sometimes it exists in a custom form that is specific to an organization, such as project management software. Sometimes it exists in a more static form, such as within a content strategy plan stored in a Google Doc or PDF.

Regardless of the form it takes, there are specific actions that your editorial calendar requires for it to guide your content strategy process. We provide brief descriptions of each of those actions now.

Turning Goals into Action Items

The first step in creating an editorial calendar is to revisit the goals of your content strategy plan and brainstorm action items from those goals. This should be relatively straightforward if you're following the process we've recommended in this chapter and your goals are Measurable, Achievable, Simple, and Task-oriented. You simply need to break the tasks mentioned in your goals into sub-tasks with proposed deadlines for completing them.

If one of your goals is to audit a specific type of content throughout your organization, for instance, then you'll need to create action items for doing so based on the advice we presented in Chapter 5, such as creating goals for your audit, conducting a content inventory, developing a rubric, assessing content, creating findings from an audit, and writing a content audit report. If another goal is to produce a specific type of content, then your first task will probably be to create a content model for that content and produce a draft of it.

In our experience, content strategists are often doing many different types of tasks at once. Very few content strategists we've spoken to are able to stop producing content in order to do a content audit, for instance. Organizations have goals of their own that must be met. Instead, many of them end up sort of building the plane while flying it: trying to create a plan as they're producing content. They may start an audit of existing content while they're still responsible for creating new content. They may have to retroactively brainstorm a content model for content they've already drafted.

The goal of a content strategy plan is for it to grow and change as content goals grow and change. This makes the planning process a whole lot messier, but also a lot more realistic. This is never more true than when collaborating with other content developers, which the vast majority of content strategists are doing. We discuss this process in Chapter 8.

Recurring versus One-off Tasks

As far as the types of action items that typically appear in editorial calendars, they are largely of two different kinds: recurring or one-off. Some tasks, such as producing new content for a faster-moving channel, such as a company blog, are recurring in that they must be completed on a regular basis. Other tasks, such as completing a content audit in order to transition content from one form to another, may have a single start and end date.

Like everything else in your editorial calendar, the duration of tasks should be driven by your goals. Ask yourself if a specific goal requires an ongoing commitment of time or if it can be satisfied through a single effort. Most of the content strategy plans we've seen have a mix of recurring and one-off tasks.

Solo versus Collaborative Tasks

Another facet of action items in your editorial calendar is whether they will be completed by individuals or groups of people. A common workflow for content development, for example, is draft, revise, edit. These tasks may be completed by a single person or a group of people. The person who drafts the content may not be the same person who revises it or edits it. Sometimes it is the same person.

As a rule, the only real benefit of content strategy planning for a single content strategist is being able to quickly replicate past tasks. Content strategy plans really shine when you have to work with other people. Imagine working with as small a group as five or six other content creators. Every time a task comes up, you, the content strategist, have to communicate the tasks, including sub-tasks, deadlines, and where to submit the content. You have to keep track of all this information, because ultimately it's your job to keep the content strategy moving forward.

Now, imagine if you have a single plan that you can refer other content developers to within your organization. You still have to follow up with them, but you save yourself lots of time and effort that you would've spent answering questions that are answered within the plan. Onboarding new content developers into your overall process is a lot easier as step one is always for them to read the plan. With an editorial calendar that has actual deadlines represented clearly, it becomes a lot easier for content developers to plan their time accordingly and you spend less time following up on wayward content.

Common Action Items

There are a lot of common tasks that you'll find yourself doing as a content strategist, just like in any profession. There are certainly going to be more than what we mention here, but some common tasks you may consider including in your editorial calendar include:

- Drafting content
- Revising content
- Reviewing content
- Inventorying content
- Auditing content
- Editing content
- Publishing/scheduling content within specific channels
- Checking analytics, audience activity, or other sources of data
- Reviewing and revising content goals
- Interviewing audience members or subject matter experts
- Meeting with other content developers or other stakeholders
- Creating or revising audience personas
- Creating or revising content models
- Learning new tools
- Updating your editorial calendar with new tasks

Checking for Alignment in Your Plan

After you create a draft of your content strategy plan, you'll of course want to review it, but what are you reviewing it for? We like to say that a solid content strategy plan should demonstrate *alignment*. And by this we mean, it should align all of its parts: goals, audiences, content channels, content models, and tasks listed within an editorial calendar. Kind of like any form of document, such as a college essay or how-to manual for building rockets, all of the parts of your plan should support one another.

This means as you review your content strategy plan, which you should do on a recurring basis even once you put it into action, you need to ask yourself questions like the following:

- Do my goals align with the audiences I've identified?
- Does it make sense that I'm going to accomplish these goals by interacting with these specific groups of people?
- Do my content channels seem appropriate, given my goals and audiences?
- Is there at least one content channel for each goal and each audience?
- Do my content models clearly align with specific goals, audiences, and channels?
- Does it make sense that I'm going to develop, publish, and distribute the types of content mentioned in each content model, given what I'm trying to accomplish with my audiences?
- Are my content models specific enough to provide clear guidelines for content development without locking me into specific channels?
- Are all tasks associated with my goals clearly broken down, accounted for, and given real deadlines within my editorial calendar?

Questions like these will help you not only develop a solid content strategy plan but are the types of questions working content strategists must constantly ask themselves as they develop, plan for, revise, and publish content.

The Content Strategy Plan Template

As we've hopefully explained in this chapter: there's a lot to content strategy planning! In a real sense, a content strategy plan is a written version of many of the processes we've described throughout this book. If you're new to content strategy, you may feel overwhelmed by this chapter as you try to envision creating a comprehensive plan for all the different processes associated with being a content strategist.

That's why we've created a handy-dandy template to get you started (Figure 7.1).

Though it certainly doesn't account for all the things a content strategist does, our template has helped scores of people, including content creators, students, researchers, and teachers, to create content strategy plans for organizations. It's meant as a quickstart guide for people who are struggling with the content strategy process and is based on our own planning process. Next we provide an activity for getting started creating a content strategy plan for an organization while making use of the template.

TABLE OF CONTENTS

Goals

Audiences

Social Media Content Model

Blog or Webpage Content Model

Email Content Model

Google Ads Content Model

Editorial Calendar / To Do List

Social Media Accounts

Website Accounts

Figure 7.1 The content strategy plan template

Source: Fillable version available here: https://bit.ly/3iBLqgy

Getting Started Guide: Creating a Content Strategy Plan for an Organization

As you can see from this chapter, creating a content strategy plan is a very involved process! How do you know how many goals to set? How do you know which audiences to include? How do you model all of the different types of content and the channels they fit in? To help get you started, we've created a Content Strategy Plan template to help you: *https://bit.ly/3iBLqgy*.

The cheat sheet includes:

- Tips for setting goals
- Tips for creating audience personas
- Tips for creating various content models
- Tips for starting an editorial calendar

Follow these steps to start your first content strategy plan:

1. Download the Content Strategy Plan template. The easiest way to do this is to go to the link (*https://bit.ly/3iBLqgy*), click File, and

then click Download. You can download the template as a Word document, an OpenDocument, or an RTF. If you're signed into your Google account, you can also go to File > Make a Copy and create a Google Doc version that you can edit directly that way.

2. Use the guidelines in this chapter as well as the tips in the template to create a draft content strategy for yourself or a local organization you work with. You might want to create a plan for promoting yourself to future employers, for example. Or maybe you volunteer with a local non-profit that is struggling to manage all of their content. Maybe you work part-time for a small business that could use this kind of strategy. Or maybe you're part of a student organization that needs to promote their volunteer activities in the local community.

3. Once you create a draft content strategy plan, be sure to show it your intended client. If it's for you, personally, show it to some of your peers to see what they think of it!

4. Once your content strategy plan draft is complete, try it out! Content strategy plans are living documents. The best way to refine them is to try using them to develop and distribute some content to see how well they work. You'll quickly learn what's missing from your plan and can revise accordingly.

References

Pew Research Center. (2021, November 23). *Demographics of social media users and adoption in the United States*. Pew Research Center: Internet, Science & Tech. Retrieved January 6, 2022, from www.pewresearch.org/internet/fact-sheet/social-media/

Rockley, A. & Cooper, C. (2012). *Managing enterprise content: A unified content strategy*. 2nd ed. New Riders.

Further Reading

Content strategy template. Moz. (n.d.). Retrieved March 30, 2022, from https://moz.com/content-strategy-template

Forsey, C. (2021, October 21). *How to develop a content strategy in 7 steps: A start-to-finish guide*. HubSpot Blog. Retrieved March 30, 2022, from https://blog.hubspot.com/marketing/content-marketing-plan

Peters, S. (2020, July 31). *10 free content strategy and editorial calendar templates*. Builtvisible. Retrieved March 30, 2022, from https://builtvisible.com/content-strategy-editorial-calendar-templates/

8 Collaborating With Other Content Developers

In the previous chapter, we discussed how you might assemble your very own content strategy plan. Now that it's time to get started developing content according to your plan, you may realize that there are some items that you can't tackle alone. Collaboration is a key element and skill set for any good content strategist and understanding the importance and value that collaboration can afford you in any given project is a must.

As we've tried to make clear throughout this book, content strategy is rarely a solo effort. In addition to having a content strategy plan that will lay out the goals, audiences, channels, content models, and essential tasks for your content, you'll want to set up a workflow for collaboration across your organization to ensure success.

Furthermore, having an effective communication strategy can aid your team's success on any given project. In fact, according to the Project Management Institute (PMI), ineffective communication can be attributed to 56% of failed projects (Monkhouse, 2015). Regardless of how you may be collaborating with others (whether in-person, virtually, remotely, or in a combination), there are many best practices to consider to ensure that you and your team communicate effectively.

Collaborative Workflows, Tools, and Best Practices

In order to set you and your collaborative partners up for success, there are a myriad of tools and best practices to consider when working on a project. Some of the tools that you may consider using to assist you and your team are: video conferencing platforms, collaborative writing tools, and peer review systems.

Note that we expand on many of these collaborative tools in Chapter 13. In this chapter, we focus on how to build collaboration and teamwork into your everyday practice as a content strategist.

DOI: 10.4324/9781003164807-8

Who Are Your Collaborators?

There are many different types of stakeholders that you may find yourself collaborating with over the course of a content strategy project. These people include, but are not limited to:

- Technical writers
- Technical editors
- Subject matter experts (SMEs)
- Marketers
- Salespeople
- Product managers
- Engineers
- Programmers
- UX designers
- UX writers

Each one of these people can provide you with information essential to your work as a content strategist, including data from previous research, information about a product, and information about audiences.

Depending on the size of the organization you're working for as a content strategist, you may be the only person with that title in your organization. At the same time, all the organizations we've encountered, from small nonprofits to large corporations, have had multiple people responsible for developing content. So while you may be one of the only designated content strategists on a given project, you are never truly on your own.

What we find to be helpful for people new to content strategy is to think about people you work with as falling within different categories of collaborator. Specifically: thinking about them as people who design content, people who market content, and people who write content.

People who design content:

- UX designers
- Engineers
- Programmers

People who market content:

- Product managers
- Marketers
- Salespeople

People who write content:

- Technical writers
- Technical editors
- UX writers

It's helpful to think about collaborators in this way so that you can clearly define the specific roles and responsibilities of each member of your team within your editorial calendar. In order to create tasks that make sense for each person involved with content, you need an understanding of what each person brings to the table, who you can go to when you have a specific question about content, and how each member may impact the workflow of a project. Which brings us to another important thing to note about collaborating: what that workflow looks like.

Building a Collaborative Workflow

While being a content strategist involves developing content on your own some of the time, often you work collaboratively to develop content, organize content, maintain or update existing content, and help other content developers create consistency across the organization. No matter what role you're fulfilling as a content strategist, however, creating a workflow for all parties involved will go a long way in contributing to your content's success.

It's also important to note that your collaborative workflow must be adaptable, much like your content. Here, we mean adaptable in that no matter who is collaborating on a particular project, they are able to provide value, contribute, and stay organized. While we provide a sample workflow for how you might structure a project here, we understand that it may look different depending on the organization, the project, and the people involved.

Our sample workflow is as follows:

- Organizing planning meetings
- Developing editorial calendars
- Editing content
- Working with subject matter experts
- Tracking content goals

So, what do these activities look like within an actual workflow? We discuss each in turn next.

Organizing Planning Meetings

Before we can start developing content, we need to come up with a plan that includes the information we discuss in Chapter 7, understand who is involved in the project, understand what information we have, and understand what additional information we need. Starting a project off with a kickoff meeting allows everybody involved with the project (from any of our categories of collaborators) to understand the role that they fill within the project.

Besides a kickoff meeting to discuss our initial plan, our goals, and what role everyone will play in the project, there are several other types of meetings that are important to schedule as the project unfolds:

- Check-in meetings to discuss the progress of the project
- Check-in meetings to discuss shifting goals or requirements
- One-on-one meetings to discuss a specific collaborator's needs

We're a fan of lean meetings during content strategy projects. Only meet when it's absolutely necessary. Remember that many content creators within modern organizations aren't doing content development full-time. If you make it too hard for them to contribute to the content strategy process, they'll often make themselves unavailable and you'll be stuck working on your own.

Developing Editorial Calendars

One thing that you may want to discuss during your planning meetings is setting up and developing an editorial calendar, which we explored in Chapter 7. In terms of collaboration, an editorial calendar allows you to channel everyone that has an impact on content toward specific deadlines and goals, which can be really helpful in a large, collaborative atmosphere.

While we discussed some of the tasks associated with an editorial calendar in Chapter 7, you may still be wondering how you physically create one during an actual project. This question has many answers, since there are many different options for creating and maintaining an editorial calendar. Here, we will provide two options that we've used frequently with clients and with our own workflows: using an online calendar and using a spreadsheet.

When using an online calendar, such as Apple's iCloud Calendar (https://www.icloud.com/calendar), Google Calendar (https://calendar.google.com/), or Microsoft's Outlook Calendar (https://outlook.office.com/calendar/), you can simply create events within the calendar. In order to assign deadlines to certain members of your team, you can simply invite them to the calendar event. You can also schedule meetings this way by attaching a meeting invitation, such as a Zoom link, to the event. Using a calendar in this way allows all members of the team to view all deadlines for which they are responsible, get reminders sent to them, and work accordingly.

On the other hand, if you want to visualize your content in a different way, a spreadsheet can work just as well. When building a spreadsheet that will serve as an editorial calendar, you want to think about what pieces of content you need to track, who is assigned to the content, what the due date is, what the status of the content is, and how content will be delivered (such as a link shared from a collaborative writing tool like Google Docs or OneNote).

	A	B	C	D	E	F
1	Title	Assigned to	Date Assigned	Due Date	Status	Link to Draft Content
2	Insert content title here	Insert name of assigned pe	Insert date this content wa	Insert due date	Not Yet	Insert link to your drafted content
3						
4						
5						
6						
7						
8						
9						
10						
11						
12						
13						
14						
15						
16						

Figure 8.1 Example editorial calendar spreadsheet

Source: Fillable version available here: https://bit.ly/3NrTgb4

Figure 8.1 presents an example spreadsheet layout based on this option.

One advantage of this method is that you can easily sort your spreadsheet by due date if you need to see what is coming up, and when—and like many calendars, most spreadsheet programs allow you to share the content with all of your collaborators.

While there are many other ways to get started with your editorial calendar, such as using custom software that your organization has developed for project management or other related tasks, in the end it's just helpful to start any way you can. You may need to revise, restructure, or switch the technology that powers your editorial calendar once you put it to use in a real project, and that's okay. You and your team will thank yourselves later for the work you put into developing an editorial calendar as you work through your project.

Revising and Editing Content

Though developing an editorial calendar that includes a schedule of all associated tasks and due dates is the best practice in content strategy, it's important to note that just because you've scheduled content to be written and delivered by a certain due date does not mean that it will be *publishable* when it gets to you. No matter how great of a content developer anyone on your team may be, there is always a need for content to go through a revision and editing process.

That process and how long it takes is completely determined by how well the content was written, which usually means how closely the content developer followed your content model (see Chapter 6). Regardless, when you are creating your editorial calendar, remember that a due date for a *draft* of content shouldn't be the same as the date the content needs to be *published.* You also need to leave time in your workflow for peer review, revision, editing, and formatting for publication.

Editing content becomes even more important when you are collaborating with other content developers who may be creating vastly different types of content and may also have different levels of ability when it comes to writing and editing. This is why there are always benchmarks that content must adhere to before it can be published. These benchmarks are often called *editorial guidelines,* or sometimes *brand standards,* or *style guides.* They include, but are not limited to:

- The overall style of content
- The purpose of content
- The target audiences of content
- The specific formatting of content
- The technologies that allow content to be delivered to specific audiences

We discuss how to edit content more thoroughly in Chapter 9, but it's important to note for our present purposes that this process needs to be a part of your workflow. Often the content strategist serves as the overall voice of the organization, meaning that once content gets to you, you're responsible for unifying it around audience needs and organizational goals. Content coming from a single organization shouldn't sound like it's coming from individuals within the organization, even though this is the reality. It should sound like it's coming from the organization itself, and this requires the work of someone who will revise and edit the content for consistency.

Working with Subject Matter Experts

As a content strategist, you are oftentimes collaborating with people who have different levels of knowledge about a particular topic. This is especially true if you are working in a technical field such as a hard science, medicine, engineering, or technology. Most content strategists gain knowledge about their core industry (e.g., pharmaceuticals) as they ·work to manage content in that industry, but there are always experts out there who have more specific knowledge that can be incredibly useful to your content development efforts. Because of this, your revision and editing process will sometimes require you to work subject matter experts, or SMEs.

This is perhaps most true in the realms of credibility, authority, and accuracy. As a content strategist, you will edit content for consistency, voice, and style. You'll make sure that that content is structured correctly according to specific content models, but you simply may not have the expertise required to determine if the information represented in the content is credible, authoritative, and accurate.

Consider the following content from a past client project, an audit of a large website:

> The TE3000-series of RF vector network analyzers have full vector measurement capability, and accurately resolves the resistive, capacitive and inductive components of a load. The user can display the vector impedance plus a range of related parameters including SWR, reflection coefficient, return loss and R–L–C equivalent circuit.

As content strategy experts, we were qualified to review this content for consistency, voice, and style, but as we are not experts in this client's core industry (radio technology), we *weren't* qualified to assess the extent to which the claims made by this content were credible, authoritative, and accurate. If the client had asked us to do so (they didn't), we would have had to recruit someone with expertise in this specific industry to review the content.

Part of your revision and editing process (and sometimes even your content development process) may require you to bring pieces of content like the above to an SME for review. This often involves asking pointed questions of the SME, such as:

- Based on your experience in the [SPECIFIC NAME OF] industry, are the claims made in this content credible, meaning believable to experts such as yourself?
- Based on your experience in the [SPECIFIC NAME OF] industry, are the claims made in this content authoritative, meaning based on the most current best practices?
- Based on your experience in the [SPECIFIC NAME OF] industry, are the claims made in this content accurate, meaning correct in all details?

Consulting with subject matter experts ensures that content coming from an organization is trustworthy, in other words, and will be seen as reliable by audiences.

As far as when to include subject matter experts in your content strategy process, we advise you to seek guidance whenever you feel ill-suited to judge the factual accuracy of a claim in content you're working on. Sometimes that guidance can simply be a quick Google search to find an article written by an expert in the field associated with the content. Other times, a subject matter expert may work within the same organization and you can simply schedule a brief meeting with them to discuss the content. And still other times, subject matter experts must be recruited from the broader field using a tool such as LinkedIn or a field-wide email listserv. If recruitment is necessary, many organizations will offer to pay the subject matter expert for their time, typically at the same hourly rate that they would a content developer within their organization.

Tracking Content Goals

Once you have determined a plan for developing, reviewing, revising, editing, and ultimately publishing your content, it's time to determine what key performance indicators (or KPIs) you will use to gauge your content's performance. This is the specific measurement you've hopefully included in your content goals for your content strategy plan, something like *we will post relevant articles to our help forum in order to decrease emails to our support team by 15% over the next quarter*. In this example, the KPI is 15%. By measuring the amount of emails the support team receives over a three-month period, you could judge if your content has the desired effect or not.

We talked extensively about content goals in Chapters 5 and 7, but the important addition here is that you need to *track* your goals over time as part of your overall workflow. Understanding what you are trying to accomplish with your content will allow you to track the overall effect of your content on a key performance indicator. However, if you never follow up on these goals to assess whether or not you have met them, then you won't know what impact your content is having, if any.

There are a wide variety of KPIs that content strategists track for organizations. They can include:

- Changes to organizational metrics such as donations, sales, or support tickets
- Unique or return visits to online content
- Bounce rate, or how many visitors find content online and then immediately abandon it
- Subscriptions to a newsletter, support system, or user help forum
- Follows or other types of engagements (i.e., likes, retweets, comments) on social media or blogs
- Readability scores, or how easy it is for audiences with specific reading comprehension levels to understand content
- Qualitative metrics such as ranking all content for audience appropriateness based on a scale of 1–10
- Ranking of content within search engine results pages (SERPs)
- Backlinks, or how many other websites link to a piece of online content

Your specific KPIs will always be unique to you, your team, and your organization. Be sure you include steps in your workflow that involve revisiting your KPIs on a semi-regular basis (such as weekly, monthly, quarterly, or yearly). Doing this kind of continual assessment will go a long way toward helping you determine what content is performing well, what content is underperforming and thus needs to be revised, and what content isn't performing at all and should be retired. Successful content strategists are always adapting and improving the way content is developed, revised, and delivered.

Getting Started Guide: Developing a Workflow

To get started putting a content strategy plan into action, review the activities associated with collaborating with other content developers from this chapter and then create a workflow for you and your team:

1. Start by defining roles for your team. Some roles to consider are: team leader, project manager, researcher, content specialist, editor, reviewer, and subject matter expert.
2. Next, discuss how work will get done in your team. Start this discussion by exploring who will tackle which tasks from your content strategy plan. Assign due dates to each task.
3. Create an online calendar, spreadsheet, or other space that you can share with your team and place all the tasks and due dates into it.
4. Discuss KPIs from your content goals and how you will track these KPIs as your project unfolds.
5. Discuss when you anticipate meeting during the course of your project and how these meetings will coincide with project deadlines.
6. Finally, discuss external stakeholders, including reviewers, editors, subject matter experts, and other members of your organization that you may need to consult with during the course of your project.

References

Monkhouse, P. (2015). *My project is failing, it is not my fault.* Project Management Institute. Retrieved March 30, 2022, from www.pmi.org/learning/library/communication-method-content-in-project-9937

Further Reading

Giordano, C. (2021, April 18). *Technical writing foundations: Mastering the art of the SME interview.* TechWhirl. Retrieved March 30, 2022, from https://techwhirl.com/technical-writing-foundations-mastering-the-art-of-the-sme-interview/

Petersen, R. (2021, October 28). *A terrific 12-step editorial guidelines template to help you establish trust with high-quality content.* CoSchedule Blog. Retrieved March 30, 2022, from https://coschedule.com/blog/editorial-guidelines-template

Sharma, P. (2022, March 15). *Why content collaboration matters and how to make a go of it.* GatherContent. Retrieved March 30, 2022, from https://gathercontent.com/blog/a-6-step-strategy-to-build-strong-content-marketing-collaboration

Six steps for working with subject matter experts for better technical writing. PerfectIt™ | Proofreading Software for Professionals. (2021, July 29). Retrieved March 30, 2022, from https://intelligentediting.com/blog/six-steps-for-working-with-subject-matter-experts-for-better-technical-writing/

Stobierski, T. (2020, September 10). *How to create editorial guidelines for your content.* Pepperland Marketing. Retrieved March 30, 2022, from www.pepperlandmarketing.com/blog/how-to-create-editorial-guidelines-for-your-marketing-team

9 Revising and Editing Genres

Even if we have the best possible strategy in place, content development is a messy process. From talking to people who are professional editors, content almost never arrives to them in perfect condition. Content developers forget requirements or fail to understand them sometimes, despite the best of intentions. Sometimes organizational goals shift during content development, or even right up until publication. And content sometimes outlives its usefulness. It has to be revised, repurposed, or retired.

For all of these reasons, it's important to have a fresh set of eyes on content. It is easier for an external reviewer to find and correct flaws, as well as to ensure that the content is the best that it can be. This is why reviewing and editing are extremely important before delivering and publishing content, and also during a process like content auditing, in which content is being reconsidered.

So, what are some of the potential flaws that can creep in? Content genres may fall short in the following areas:

- Content
- People
- Reasons
- Time

(Casey, 2017)

When revising and editing specific genres of content, it's thus important to consider these areas as important benchmarks.

Content

The first benchmark for content is that it needs to be understandable to your audience. This may sound like a basic aspect, but it is the most important aspect of content because if the audience can't understand what message your content is trying to convey, it is essentially useless. This can be very disheartening based on the time and effort that may have been spent on developing that content.

DOI: 10.4324/9781003164807-9

For example, let's say we are tasked with creating flyers to invite Spanish-speaking families in a local community to an informative workshop on how to register their children for preschool for the upcoming school year. It would be extremely important to make sure that the flyer is correctly translated into Spanish. If the translation is incorrect or incomprehensible, the families may not understand the purpose of the workshop or when the workshop will take place. Additionally, it is also important to make sure the dialect chosen reflects the dialect of the Spanish-speaking families in the local community. For example, there are enough variations between the Spanish spoken in Spain and the Spanish spoken in Guatemala that these differences in dialect could affect the audience's understanding of the message. The general idea would be present, but some of the smaller points and nuances of the flyer's message would be lost.

In addition to understanding content, your audience must also be able to use the content to achieve their goals. If the audience can't use the content, they may be confused as to why they are receiving it or may fail to achieve the goals that require the content.

Using the previous example, what would happen if flyers for the preschool registration workshop were not delivered to families with small children, but were accidentally delivered to senior citizens that do not have any grandchildren living in the same community? Those senior citizens would not have any personal use for the workshop. Also, not knowing any small children in the community, the senior citizens would not be able to share the information about the upcoming workshop with anyone who may benefit from attending.

This is an example of why it is important to review and edit content to make sure it is understandable and useful to your audience.

People

Content can also fail if it isn't catered to the right audience. This should be a conscious decision from the planning stages and should continue through publication and delivery. Otherwise, there is no way to ensure your content reaches the real, live people who need it.

Using the previous example of the preschool registration workshop, the right audience would be Spanish-speaking families with small children in the local community. If we know this information, we can make sure the flyers are developed, published, and delivered to these families. Also, if we know the number of families in the community, we can evaluate the effectiveness and usefulness of the flyers by tallying how many families attend the workshop.

If we don't identify the right audience for the preschool registration workshop flyers from the outset of the content planning process, we may not send the flyers to the correct households that could use the information. Additionally, not knowing the correct number of families to send the flyers to could result in sending flyers to every household in the community, which would be a waste of resources, or failing to budget for enough flyers to be printed and sent, which would result in some families being left out of this opportunity. If a non-profit is the sponsoring organization for the flyers, the printing and mailing of a much larger volume of flyers would dramatically increase the cost of promoting this workshop and possibly overly tax the organization. On the other hand, failing to deliver sufficient flyers may hurt the organization's bottom line when their outreach goals for the year aren't met.

These are examples of some of the reasons why is it important to be deliberate in your decision to target specific audiences, which are composed of groups of real, live people with concrete needs.

Reasons

It is also important to determine why you are creating specific content. All content you develop should have a purpose, otherwise there is no reason to create that content in the first place. Again, this may seem obvious, but you wouldn't believe how many organizations we've worked with over the years who have churned out a wide variety of content genres with no clear rationale for doing so.

Using our running example, let's assume that we were given the vaguer task by a client organization of "creating a flyer for a preschool thing." From this task, it isn't clear why we are creating this flyer. Is this an informative flyer? Are we persuading an audience? Is the flyer serving as a reminder? Also, if the only information we have about the event being described is that it's "a preschool thing," it's not clear what type of event we are promoting or what information we should include on the flyer. Finally, it is not clear who the audience for this "preschool thing" would be. Is it directed toward educators? Parents? School board members?

Not having clear reasons for content makes it difficult to understand why content should be created, let alone what information that content should entail, how that information should be conveyed, and who the content should target.

Time

Finally, it's important to know when to deliver content. If content is delivered too early, your audience may forget about the message. If content is delivered too late, your audience may not be able to do anything with the information it contains.

Using our running example, the timing of the flyers' delivery to households is extremely important. If the flyers are delivered before a given family has any eligible children, it is very unlikely that the family will attend the workshop and may not remember about the workshop in a couple years' time. On the other hand, if another given family who has eligible children receives the flyer after the registration deadline has already passed, it will be too late for that family to register their child for preschool, let alone attend an informational workshop about the process of registration.

There is a specific window of time when receiving the workshop flyer would be useful, in other words. Delivering the flyer too early or too late will not only negate the flyer's use, but will also be a waste of time and resources for both content developers and audience members. This is why timing is so important.

Related to this, errors in the flyer that are discovered late into the content process may delay the delivery timeline. This is why it's important to set aside sufficient time before an intended delivery date to review and edit content.

A flaw with any one of these areas can cause genres to fail. It's much easier to correct flaws in the pre-publication phase of most genres than it is to have to roll back flawed content that is already published, and then to create, edit, and publish revised content in a very quick time frame. Taking a little extra time to make sure content is correct before publishing through regular reviews will save time and potential frustration for everyone involved.

Reviewing

As mentioned in the previous sections, it's important to build extra time into a given genre's timeline to ensure that it meets the requirements of content, people, reasons, and time. This means making sure a genre is in its best possible state before publication. Reviewing is *the process of examining content with the intent of making changes if necessary*. In nearly every case, content will need revision. Content genres are complex. Even the simplest ones entail a host of requirements, processes, audience needs, and design challenges. This is why it's essential to review all content before it gets to an audience.

Although many content developers, in our experience, are used to reviewing right before delivery, reviewing can also happen at any step in the content creation process. And we would argue that reviewing should start early on in the content creation process.

Using the earlier example of the flyer for a preschool registration workshop, if the flyer was reviewed early on in the creation process, there would be fewer potential changes to make later on in the process, which would mean less time editing overall. Additionally, an early review would help course correct the content creator and prevent the likelihood of continued errors related to the potential changes. Conversely, if the same flyer wasn't reviewed until it was about to be sent to families, there would a greater potential for errors, which would then push it back to editing and possibly even development, which would add time to overall production.

Also, at this point, it might be too late for any course correction because the flyer is already completed and locked into a specific publication format, such as proofs that have been sent to a company that provides custom mailers. The additional delay to make these potential changes could affect the delivery of the flyer, which could lead to it failing the benchmark of time.

But it's just a flyer! You may be thinking. And in our experience, relatively simple content genres like these are often the most underestimated by organizations. Remember: the broader the audience, the more impact a single content genre has. If this flyer was advertising one of the main programs that our example non-profit offers, then complications with its production could have real impact on the organization. Most non-profits have to measure their impact and report this impact to several agencies each year, including at the state and federal level, depending on the nature of the non-profit. These reports can have an effect on future funding decisions from external agencies, including the ability of a non-profit to apply for future grants.

Let's look at this another way: we've never encountered a single organization that only focuses on one simple genre of content. Even small organizations are now rife with content: not only flyers, but webpages, blogs, emails, social media posts, and all the other genres we've discussed throughout this book. Magnify the issues we've discussed in the context of a single genre across an entire organization and you have a real nightmare on your hands. Are the social media posts being reviewed before being published? Is all the information on the blog post advertising the latest fundraiser accurate? Is there an internal content repository that staff in the organization can use to get their work done, or are they wasting countless hours developing content from scratch every time?

And in our experience, the smaller the organization, the fewer staff and budget there is for this kind of systematic review, and thus the more mistakes that are made.

Due to these factors, it's our suggestion that every organization use a combination of reviewing strategies in order to make sure content is going in the right direction throughout the creation process, as well as reviewing the content again before delivery. This will reduce additional frustrations and extra work needed down the line. Also, with a layered review process there's a higher likelihood that content will be effective, technically correct, and delivered on time.

Some of the different layers of review that happen to content within a successful organization are:

- Formative review: reviews of initial drafts of genres to make sure they are headed in the right direction
- Summative review: reviews of semi-finished drafts of genres to make sure they meet basic quality standards
- Editorial review: corrective reviews that ensure content perfectly matches organizational guidelines for tone, style, audience appropriateness, and grammar

And during a review process, there are two different categories of changes to content that can be made: content editing and copy editing. In the closing section of this chapter, we explain the difference between the two. In order to have an effective review process, both types of editing should be used to ensure the best possible content is delivered to your audience.

Content Editing

When we hear the word *editing*, we may think about correcting, condensing, and making other changes to a piece of content before publication. Content editing is very similar to this notion of editing, but it has a very strong focus on being understandable. We can define content editing as "editing with the goal of reviewing writing to check how effective, cohesive, and understandable it is" (Clark-Keane, 2019). When doing content editing, you will make changes to content to ensure it is in line with your organizational goals as well as the goals and needs of the content's specific audience. This can include checking that it meets the benchmarks we mentioned earlier:

- Content: the correct content was created, content is in the voice of your organization, and is using consistent terminology
- People: content is clear, understandable to your audience, and has a logical progression
- Reasons: content accomplishes its intended purpose effectively
- Time: the content is timely, meaning current and authoritative

Once content editing is completed, we can move on to the next type of editing: copy editing.

Copy Editing

While content editing focuses on the clarity of the content, how it will be perceived by the audience, and the message of the content, copy editing looks for smaller errors in the content itself. This can include:

- Spelling
- Grammar
- Punctuation
- Syntax
- Style guide requirements
- Image content and placement
- Meta descriptions and captions

(Clark-Keane, 2019)

There are many great books on copy editing out there, but we recommend *The Copyeditor's handbook: A guide for book publishing and corporate communications* (4th edition). It's a wonderful, no-nonsense book with many, many resources on copy editing. Everyone dealing with content directly should have a copy of it.

The bottom line is that content is an important product within an organization. And typically, it needs to be reviewed multiple times to ensure your audience receives effective content.

Getting Started Guide: Organizing a Peer Review Circle

One of the best ways to make sure content is reviewed effectively is to organize a peer review circle, or a *group of people who agree to review content at any stage of its development process*. Anyone can organize a peer review circle. And it's relatively simple to do so:

1. Gather some of your peers who care about content and who want to improve it.
2. Discuss the guidelines for your peer review circle. What are you going to be doing? Formative review? Summative review? Editorial review? Some combination?
3. Discuss the benchmarks you'll be reviewing for: content? Reasons? People? Time? Something else?
4. Start reviewing content using the process you've developed and get the feedback to the content developers who created it!

Like editing, peer review is a skill that must be honed over time. The most important things to do when you're starting out are to decide on the benchmarks that content must meet and rigorously apply those. This may require some learning on your part. If you're doing copy editing for the first time, you're going to have to school yourself in how to do that.

References

Casey, M. (2017, March 30). *Why you need content strategy before editorial planning.* Content Marketing Institute. https://contentmarketinginstitute.com/2017/03/content-strategy-editorial-planning/

Clark-Keane, C. (2019, November 6). *The last guide to content editing you'll ever need.* WordStream. www.wordstream.com/blog/ws/2019/11/06/content-editing

Further Reading

Albers, M. & Flanagan, S. (2019). *Editing in the modern classroom.* Routledge.

Büky, E., Schwartz, M., & Einsohn, A. (2019). *The copyeditor's workbook: Exercises and tips for honing your editorial judgment.* University of California Press.

Einsohn, A. & Schwartz, M. (2019). *The copyeditor's handbook: A guide for book publishing and corporate communications.* 4th ed. University of California Press.

How to start a peer review circle to improve your content. Content Garden. (2018, October 3). Retrieved April 2, 2022, from www.contentgarden.org/peer-review/

TeamKapost. (n.d.). *Why every content strategy needs a managing editor.* Upland Software. https://uplandsoftware.com/kapost/resources/blog/why-you-need-a-managing-editor/

10 Ensuring Content Usability and Accessibility

Why Is Content Usability Important?

While revising and editing your content is crucial in making sure that your content is error-free, concise, and understandable to your audience, there are other ways to test your content for its impact and usefulness for a given audience. Two of those ways with which all content strategists should be familiar are usability and accessibility.

Defining Usability

According to the Interaction Design Foundation, usability is "a measure of how well a specific user in a specific context can use a product/design to achieve a defined goal effectively, efficiently and satisfactorily" (*What is usability?*). We would add content to this definition: usability is also important when it comes to content. If audience members aren't able to use your content to achieve their goals and deal with their challenges, effectively and efficiently, then your content isn't usable.

Before diving into usability testing and explaining the different methods for testing content usability, it's important to understand the overall method of usability testing.

Differences Between Traditional Usability Testing and Testing Content

Usability testing is essentially the process of recruiting users to test out a product. This testing process involves a UX practitioner asking users to attempt a series of tasks and then observing how well they are able to complete the tasks. There is also an interview portion where the UXer asks users to explain why they completed the tasks the way they did.

Usability testing gives UXers insight into what the user is thinking when they use a product, which helps them improve the product so that its features

DOI: 10.4324/9781003164807-10

match the user's expectations. The benchmarks for whether a product is usable vary, but often include quality components such as:

1. Learnability: How easy is it for users to accomplish basic tasks the first time they encounter the design?
2. Efficiency: Once users have learned the design, how quickly can they perform tasks?
3. Memorability: When users return to the design after a period of not using it, how easily can they reestablish proficiency?
4. Errors: How many errors do users make, how severe are these errors, and how easily can they recover from the errors?
5. Satisfaction: How pleasant is it to use the design?

(Nielsen, 2012)

The focus of content-focused usability testing is much more specific, however. We're focused on methods that can help us test our wording, messaging, our users' reading comprehension level, and how well our content assists our users in accomplishing what we need them to accomplish. Although this chapter and its methods list are not exhaustive, we have identified some of our favorites and some of the most helpful research methods to have at your disposal as a content strategist.

Content Usability Methods

When searching through the expansive field of user experience design (UX) and the possible methods that you could utilize for testing products, designs, or content, it might feel overwhelming figuring out where to start. In an effort to ease you into the world of content usability, we have selected some of our favorites that can give you a variety of options for testing, whether you're testing on your own or with a team, or bringing in real users to test your content.

Content Usability on Your Own

If you're looking for usability methods that you can use before recruiting users to test your content, we have a method for that. Sometimes it's helpful to assess your content on your own before inviting users to help you assess it.

Competitive Analysis

We discussed doing competitive analyses in Chapter 4 in the context of identifying content types and channels, and in Chapter 6 in the context of building content models. You can also do a competitive analysis for usability, however.

According to Schade (2013):

> Competitive usability evaluations are a method to determine how your site performs in relation to your competitors' sites. The comparison can be holistic, ranking sites by some overall site-usability metrics, or it can be more focused, comparing features, content, or design elements across sites.

From this definition, we want to focus on two things. First, in the context of content strategy, we are identifying usability metrics associated with content, rather than for a design, though design can be part of a content audit, as we described in Chapter 5. Second, when identifying competitors, we are looking for content that is trying to do the same things our content is trying to do, meaning these "competitors" may simply be organizations that are similar to ours rather than ones that directly compete with us.

To begin a competitive analysis, you need to identify content creators who are developing content that is similar to the type of content you are currently creating or to content that you want to create in the future. In this sense, you want to find organizations who have content that makes you say "wow, I wish we'd thought of that!" or "this is the kind of content we need to be creating!" As a rule, you're trying to compare to leaders, not followers, in other words. You want to improve the usability of your content, not compare down to organizations that you feel like have less usable content.

There are several different ways that you can go about organizing this information as you collect it, but we recommend (you guessed it) a spreadsheet (Figure 10.1).

If this spreadsheet looks familiar, that's because it's the same spreadsheet we recommended in Chapter 5 in the context of content auditing. In fact, it's arguable that a competitor analysis is a kind of content audit where the goal is to examine the content of other organizations. Just like a content audit, you want to identify criteria for determining what's effective in what you see.

	A	B	C	D	E	F
1	Content Source	Content Type	Content Channel	Date Published	Assessment Criteria #1	Assessment Criteria #2
2						
3						
4						
5						
6						
7						
8						
9						
10						
11						
12						
13						
14						
15						
16						

Figure 10.1 Example content audit spreadsheet

Source: Fillable version available here: https://bit.ly/30jUQaT

Once you've identified the other organizations you feel have great content, you need to identify what makes it great. You can look back at Chapter 5 to see some of the assessment criteria we mention there. In general, however, besides recording the basic information such as the content source, the content type, the content channel, and the date the content was published, you want to think about what you want your content to be able to do. What are your criteria for a good piece of content? Is it specific metrics such as likes, shares, retweets, or views? Is it being well written? Well designed?

Remember that your focus here is usability. So, no matter what criteria you choose to use, your ultimate goal is to find content that you think *is usable for an audience*, meaning content that is effective at helping audience members achieve their goals. Once you've identified several examples (we recommend 2–3) of such content, you then want to compare and contrast your own content to this competitor content.

If you've decided that what is effective about another organization's webpages is that they include clear, effective calls to action on every page, for instance, once you've collected 2–3 calls to action, collect some of your own calls to action and compare them. What is effective about the competitor calls to action? How effective are your calls to action, in comparison?

Don't be afraid to do some research into best practices as well. Finding some articles by thought leaders on what makes a given content type effective can only make your competitor analysis easier as it will help you clarify the criteria you're using.

Usability Testing Content With Users

Though data collected through competitive analyses is no doubt valuable, data garnered from usability testing with real, live users is simply invaluable. Having users sit down and attempt to accomplish their goals while using your content will tell you more about the usability of your content than any other method.

Based on best practices for usability testing, testing content is relatively straightforward:

1. Create tasks that align with the goals of the content you want to test.
2. Recruit 3–5 test users who match the demographics of your target audience.
3. Ask the users to attempt the tasks and observe them doing so.
4. After each task, ask the users follow-up questions about the task.

What you're trying to assess here is how easy it is for users to accomplish specific goals while using your content. To simulate this, you need to break down the intended goals of the content into discrete tasks that are stated as commands. These commands should be open-ended, information-seeking tasks, such as the following:

- Finding information to help them with a problem they're having
- Researching a product or service to see if it is a good fit for their current needs
- Navigating a complex piece of content to find specific information
- Searching for information that is relevant to them in a database of articles

These tasks should be clear extensions of your content goals; in other words, they should align with what your content is trying to do. That's what you're assessing: how effective is your content at enabling people to use it in specific ways?

People tend to use content in all sorts of ways, meaning that the goals you have for your content may not align with the goals your users have. This is why, unlike traditional usability testing where you want to give very specific tasks such as "register for the next train to arrive," when usability testing content, you want more open-ended tasks. You want to see how a user tries to use content to achieve their goals, which will also tell you a little bit more about what their goals are.

Sometimes users don't even know what their specific goals are when they start consuming content. Think about your own experiences with content: how often have you started looking for one specific piece of information only to discover that your entire reason for looking for that information wasn't actually suitable to your overall problem? How many times have you been doing research online for an article or school-based assignment only to realize your topic was far more complex than you had initially considered?

For more details on doing usability testing of content with actual users, we highly recommend the article by Moran (2021) that delves into a lot of the challenges and best practices of this method.

A/B Testing

A/B testing is one of the most common usability methods for testing products, designs, and content. The reason for this is that A/B testing is primarily a comparison method, used in order to compare, for our purposes, two or more pieces of content to one another in order to determine which one is more effective.

Essentially, A/B testing puts Content Piece A against Content Piece B to see which one performs better when presented to users. For example, if you were writing a revised version of a set of instructions for a how-to guide that was 5 years old, you might designate the old content as Content Piece A and the new content that has been revised and updated as Content Piece B and have different users perform the same tasks using each piece of content.

Using this method, you can determine which piece of content is more usable by comparing the performances of each group of users.

It should also be noted that although the method is called A/B testing and is most commonly used to compare two things against one another, there is always the opportunity to test more than that. There is no real limit to how many versions of a piece of content you can test with this method. However, you should be warned that the more pieces of content you choose to compare against one another, the more difficult it will be to analyze which ones are truly superior. You will also need to recruit 3–5 users *per version* to get significant data on each piece of content in order to make a clear case as to which piece of content is better and why, which can add significantly to your workload.

This method can be very helpful to get started on a revision to a piece of content, to test if a new version of content is headed in the right direction, or to help an indecisive team make a decision on a piece of content.

Impression Tests or Five-Second Tests

Another method that can be helpful when usability testing content is what is known as an *impression test* or a *5-second test*. As its name suggests, this test involves showing pieces of content to users for a very short period of time (perhaps 5 seconds, perhaps just a little bit longer depending on the amount of content), in order to garner their first impressions.

For example, maybe you want to assess whether the copy on your organization's homepage is easy to understand for a first-time visitor. You could have test users look at the homepage for 5 seconds, and then ask them to tell you what they took away from the content. Could they get the gist of it quickly? Could they immediately understand the purpose of the page? Was the font easy read? What did they feel was highlighted the most?

Or maybe you're testing to ensure that users can see a call-to-action button right away on a subscription site. Doing a 5-second test might allow you to determine whether or not your button stands out from the rest of the content. Were the users' eyes immediately drawn to it? If not, what were they drawn to?

The feedback that you get from test users while using this method is best catered toward targeted, obvious elements of content, such as:

- Images
- Color schemes
- Font sizes
- Call-to-action buttons
- Callout sections
- Warnings
- Important notices

If you really need users to notice something quickly in a piece of content, this is a great way to assess if you've accomplished that or not.

Cloze Tests

Lastly, we want to discuss what is known as a *Cloze test*. This is a relatively new concept for testing longer chunks of content, typically somewhere between 150 and 250 words. To conduct this kind of test, you begin by removing every 5th word from a piece of content. Then, you show it to your user and ask them to read through the content and fill in the blanks with what they think the missing words are.

For example, a sentence from a Cloze test might look a little something like this:

> *If you want to _____ out whether your site _____ understand your content, you _____ test it with them.*
>
> (Colter, 2010, emphasis original)

The rationale behind this method is that if your content is effective, your users should be able to fill in the blanks with words that are close to the original words you chose. Your target is that users should be able to get the original words around 60% of the time.

If your users can get above the 60% mark, your content is pretty well written and understandable *even* with the 5th word removed. However, if your users fail to fill in your blanks with the correct word 60% of the time or more, you need to think about revising your content, and how you can make it more understandable for your audience.

While these are not the only usability testing methods out there or the only ways to test your content, they are some of our favorite methods for gathering data on what's working, what's not working, and how you can make your content more effective for your users.

Accessibility Guidelines

In addition to ensuring your content is usable, you also want to make sure that your content is *accessible*. Accessibility is a key benchmark for all content, but especially for content posted to the web where a device must be used to access it, which can create significant complications for some users.

Defining Accessibility for the Web

According to the Web Accessibility Initiative (W3C):

> Web accessibility means that websites, tools, and technologies are designed and developed so that people with disabilities can use them. More specifically, people can:
>
> - perceive, understand, navigate, and interact with the Web
> - contribute to the Web
>
> (Henry, 2022)

Just like usability, from this initial definition, accessibility may not seem like it applies specifically to your content. However, Web Content Accessibility Guidelines (WCAG) expands on this W3C definition by adding:

> Web "content" generally refers to the information in a web page or web application, including:
>
> - natural information such as text, images, and sounds
> - code or markup that defines the structure, presentation, etc.
>
> (Henry, 2022)

As content strategists, it is absolutely essential that our content is accessible to all audience members, including those with disabilities. Creating accessible content is always a process and one of the best ways to ensure our content is accessible is to recruit test users who have disabilities as part of our regular usability testing.

Next we delve into some other effective ways to make content more accessible, however.

How to Create Accessible Content

Part of the process of creating accessible content is understanding how we can create content that is accessible from the get-go. And that involves understanding how users with disabilities consume content.

One of the most popular accessibility tools used by web users is a screen reader. Screen readers allow users who have sight-related issues to navigate through web content. Most screen readers facilitate this process in one of two ways:

- Text-to-Speech, where the words are read directly to the user, or
- Braille display, where the user can read the output through a tactile pad that translates the words into Braille.

(*How screen readers*)

To help users who use assistive technologies like screen readers, you should be adding the following elements to your content:

- Headings (e.g., H1, H2, H3) to all of your content
- Alt text to your images
- Captions to all of your videos
- Transcriptions to all of your audio content

Headings

When developing written content that will be displayed on the web, such as that contained in blog posts and web pages, it's important to utilize different heading sizes so that users utilizing a screen reader can better understand how the content is organized. For example, a level-1 heading, or a <h1> tag, can help to indicate the largest heading. A level-2 heading would be a sub-heading or sub-topic underneath that level-1 heading, and so on.

Creating this hierarchy of headings through your content is actually helpful to all users, however, because it clearly demonstrates how content flows from one section to the next.

Alt Text

Alt text, or alternative text, is another important component to add to your content. Alternative text is text attached to images that are displayed on the web that describe what the image is for users who can't see it. According to WebAIM, alternative text serves many different, but important, functions:

- Screen readers announce alternative text in place of images, helping users with visual or certain cognitive disabilities perceive the content and function of the images.
- If an image fails to load or the user has blocked images, the browser will present the alternative text visually in place of the image.
- Search engines use alternative text and factor it into their assessment of the page purpose and content.

(Alternative text)

So, not only can having well-written alt text help the visually impaired understand images in your content, but it can also help with your content's SEO as well!

Captions and Transcriptions

In addition to headings and alt text, which primarily help make written content and static images more accessible, it's also important to include captions for video and transcriptions for audio.

A lot of services that enable you to host video and audio content online auto-generate captions and transcripts for you, including popular tools like Zoom and YouTube. However, this auto-generated content is not always as accurate as it could be. While these tools are helpful to get you started, you will want to make sure that you edit all of this content for accuracy.

How to Adapt Older Content

While creating usable and accessible content is a great place to start, it doesn't make up for the fact that there may be a lot of old content within your organization that is not usable or accessible. It is in the best interest of all organizations to ensure that *all* content is usable and accessible, however, which is why we close this chapter with some tips for adapting older content for usability and accessibility.

How to Adapt Older Content for Usability

There's no simple solution for adapting older content for usability. Like any content, you want to assess it using the methods we've presented in this chapter and then adapt it accordingly. However, the first question you should ask yourself is: what older content is still necessary and what can be retired? As part of a content inventory and audit (see Chapter 5), an important step to ensuring all content is usable is to simply retire older content that isn't.

When you audit older content and determine that it isn't usable but is still necessary for certain segments of your target audience, then you'll need to begin the process of testing, analyzing, and revising all this content. This can be a real nightmare for an organization, to be completely honest, which is why many organizations we've encountered haven't done it. We've heard horror stories of organizations with thousands (and even tens of thousands) of pages of content that have never been formally assessed.

Though the work involved in auditing older content is daunting, we ask organizations we work with: what is the alternative? To be weighed down by ineffective content until the end of time? To continue to experience the same content-related problems, the same frustrated customers and internal stakeholders?

To make things as simple as possible, when auditing older content, you'll probably want to add criteria like the following to your audit spreadsheet:

1. date of last access (i.e., the last time an audience member viewed this content)
2. aligns with current organizational goal (yes/no)
3. aligns with current audience goal (yes/no)

Though this won't tell you how to adapt older content, it can at least allow you to quickly assess each piece of older content in a systematic fashion. In our personal experience, many organizations are carrying around oodles of older content that hasn't been accessed in a long time, that doesn't align with their current goals, and that doesn't align with their audience's current goals. When content fails this simple test, it's probably time to retire it, or at least store it away from public consumption until such time as it might be needed.

How to Adapt Older Content for Accessibility

Similarly, when auditing older content, it's important to assess it for accessibility issues as well. If you find after reading this chapter that a lot of what we've discussed is new to you, then chances are the content you've developed in the past is not accessible. This means that you're going to need to go back and revise that content to make it accessible.

And believe us when we say that if the consequences of not having usable content are sizable for organizations, the consequences of not having accessible content can be even more severe! We all should want our content to be easily accessed by all audience members, regardless of disability status. But remember, and rightfully so: people with disabilities are a protected class within our society, meaning that they have the right to be presented with content that they can make use of, despite any impairments they may experience. Accessibility guidelines used to only apply to government websites, which are legally required to ensure all content is accessible, but as several high-profile lawsuits have proven, no organization is truly immune from these requirements (Chartier, 2008).

Like usability, the process of adapting older content to meet accessibility guidelines starts with the simple test we described in the last section. If older content isn't being accessed and doesn't align with current goals, get rid of it! However, if older content must stay and isn't meeting current accessibility guidelines, then it will need to be adapted to meet the guidelines we've described in this chapter.

Getting Started Guide: Creating a Plan to Assess Content for Usability and Accessibility

One of the best things you can do to ensure your content is usable and accessible is create a plan to review all of your content for these criteria. This includes auditing older content that currently exists within an organization as well as making sure you review all new content that gets produced. This plan should go right into your content strategy plan, so you make sure it happens. Here are some tips for drafting such a plan:

1. Add the following columns to your content audit spreadsheet if you've already started one, or create a new spreadsheet for this purpose: (a) date of last access (i.e., the last time an audience member viewed this content), (b) aligns with current organizational goal (yes/no), (c) aligns with current audience goal (yes/no)
2. Draft a plan for assessing all your content with these criteria. Who will help you accomplish this? Who in your organization can you trust with this task? If there's a significant amount of content, don't try to accomplish this task alone.

3. Next, review the criteria for usable and accessible content from this chapter. From these criteria, draft some assessment criteria and add these to your audit spreadsheet. This way, you can assess any content that you don't plan to retire as soon as your criteria in the aforementioned 1. indicate that you're probably not going to retire that content. That way you don't have to take a second pass to know which content needs to be adapted.

4. Review the methods for assessing content usability and accessibility in this chapter. Which methods seem most appropriate for assessing your content? Add these methods to your draft plan.

5. Put your plan into action!

References

Alternative text. WebAIM. (n.d.). Retrieved April 2, 2022, from https://webaim.org/techniques/alttext/

Chartier, D. (2008, August 28). *Target to pay $6 million to settle site accessibility suit*. Ars Technica. Retrieved April 2, 2022, from https://arstechnica.com/uncategorized/2008/08/target-to-pay-6-million-to-settle-site-accessibility-suit/

Colter, A. (2010, December 14). *Testing content*. A List Apart. Retrieved April 2, 2022, from https://alistapart.com/article/testing-content/

Henry, S. (Ed.). (2022, March). *Introduction to web accessibility*. Web Accessibility Initiative (WAI). Retrieved April 2, 2022, from www.w3.org/WAI/fundamentals/accessibility-intro/

How screen readers make digital content accessible. AudioEye. (2019, April 5). Retrieved April 2, 2022, from www.audioeye.com/post/what-is-a-screen-reader

Moran, K. (2021, February 7). *Testing content with users*. Nielsen Norman Group. Retrieved April 2, 2022, from www.nngroup.com/articles/testing-content-websites/

Nielsen, J. (2012, January 3). *Usability 101: Introduction to usability*. Nielsen Norman Group. Retrieved April 2, 2022, from www.nngroup.com/articles/usability-101-introduction-to-usability/

Schade, A. (2013, December 15). *Competitive usability evaluations: Definition*. Nielsen Norman Group. Retrieved April 2, 2022, from www.nngroup.com/articles/competitive-usability-evaluations/

What is usability? The Interaction Design Foundation. (n.d.). Retrieved April 2, 2022, from www.interaction-design.org/literature/topics/usability

Further Reading

Colman, G. (2016, May 17). *The Writer's Guide to making accessible web content*. Zapier. Retrieved April 2, 2022, from https://zapier.com/blog/accessible-web-content/

Farmen, N. (2019, September 3). *A/B testing: Optimizing the UX*. Usability Geek. Retrieved April 2, 2022, from https://usabilitygeek.com/a-b-testing-optimizing-the-ux/

Horton, S. & Quesenbery, W. (2013). *A web for everyone: Designing accessible user experiences*. Rosenfeld media.

Impression testing. Impression Testing | Usability & Web Accessibility. (n.d.). Retrieved April 2, 2022, from https://usability.yale.edu/usability-testing/impression-testing

McGee, L. (2021, January). *Accessibility.* W3C. Retrieved April 2, 2022, from www.w3.org/standards/webdesign/accessibility

White, K., Abou-Zahra, S., & Henry, S. L. (Eds.). (2020, December 1). *Writing for web accessibility—tips for getting started.* Web Accessibility Initiative (WAI). Retrieved April 2, 2022, from www.w3.org/WAI/tips/writing/

WishDesk. (2019, July 2). *How to create accessible content: 10 useful tips.* WishDesk. Retrieved April 2, 2022, from https://wishdesk.com/blog/how-to-create-accessible-content

11 Delivering, Governing, and Maintaining Genres

One mistake we see organizations make all the time is publishing content and then never returning to it. Content develops a life of its own once it leaves an organization and gets delivered to a channel and the audience attached to that channel. This has never been more true as channels multiply. Older readers of this book can remember when there were only a few channels organizations had to worry about. Now there are countless.

And several of these channels archive content for all the world to see for an indefinite amount of time. You may think that the Facebook post you published six years ago is gone and forgotten but then it resurfaces. Suddenly a website visitor finds a blog post from last year advertising a special offer that you're now forced to extend to them because you never retired the post or indicated an expiration date. A frustrated consumer finds specifications for an outdated product on your help forum, spends hours trying to apply invalid updates to their software, and emails your support team demanding that they refund their money for the inconvenience.

You can no longer assume that once you put content into the world that it will just go away. Rather, you need to plan that it will stay with you unless you purposefully retire it. This means auditing published content on an ongoing basis to ensure it's still relevant. It also means reusing content when appropriate. But mostly, it means having a solid plan in place for doing both of these things, so you don't end up alienating your audience with outdated, irrelevant content.

Content Governance: Auditing Content Genres and Maintaining Content Relevance

The tried-and-true method of checking on published content to assess it for continued relevance is the content audit, which we discussed thoroughly in Chapter 5. The common process we described there was explained as occurring at the beginning of the content strategy process: we're getting to know new content, perhaps a new organization. Perhaps we've been asked to review older content by an organization who's concerned with some bottom-line issue, such as a decline in website visits. Regardless, we need to get an understanding

DOI: 10.4324/9781003164807-11

as to the state of current content within an organization, so we do a content audit, maybe the first one the organization has ever commissioned.

The process of auditing isn't one and done, however. Rather, content needs to be reevaluated whenever one of the following things happen:

- it becomes outdated
- organizational goals that directly affect content shift
- tools used to develop or maintain content are retired
- audience needs shift

You can probably see the problem already: in our experience, organizations are rarely on the ball when these things happen. Actually, most organizations that come to us for a consultation are in dire straits. Their content has been flagged by Google as not mobile-responsive and now they've lost nearly all of their spots in search engine results. They have tons of content in a single technology that is now no longer supported. A new initiative in their organization requires them to rethink all of their content, but they have never done an audit, or if they have, they haven't maintained that process.

And, unfortunately, with all the various technologies people use nowadays to publish content, there isn't a single application out there that will automatically alert you when your content is outdated. Website content management systems (CMSs) don't come bundled with a "remind me to audit this blog post" feature. Social media platforms don't automatically warn you that the event you posted in 2013 is still active. Email newsletter software doesn't come with a "remember to update your audience list" reminder.

So, your content is going to become outdated at some point. Expect that. Here's what you can do about it:

- Store content in an authoritative repository
- Publish content from this repository on demand rather than starting from scratch each time
- Use content models to assure yourself that content that is published meets current organizational and audience needs
- Add an ongoing content audit to your content strategy plan, at least yearly
- Track metrics to ensure you're on the right track

This process is commonly referred to as *content governance*.

As you can see from Figure 11.1, content governance is an ongoing process. It doesn't end when content is published. And should begin as soon as you start planning content. That's the only way to ensure the process actually gets off the ground.

If content governance sounds like a lot of work: there's a simple solution you may not have realized. When you did that initial content audit, if

Content Governance Process

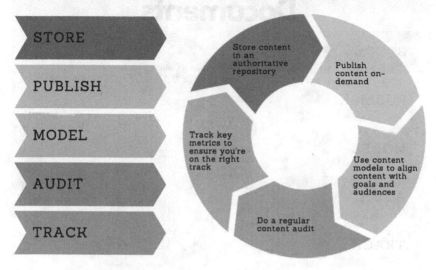

Figure 11.1 Content governance process

you took our sage advice, you created a spreadsheet to track all the content you audited. Guess what? That can now become part of your authoritative repository. You have all the content linked to there, at least, so it's a start.

Now, every time you go to publish content, you should consult your spreadsheet and add that new content to the spreadsheet at the same time you publish it to a channel. We're already on step 2! The content models should already exist in your content strategy plan, so there's step 3. When you do your next audit, it should be much easier because you can just update your existing spreadsheet (step 4). And your metrics should be part of your content goals from your content strategy plan.

You've already got the tools to make content governance a reality at this point (Figure 11.2).

This process also encourages *content reuse*, which is a key process that many organizations ignore. We turn to that next.

Like all the best practices in this book, content governance can be done in a lot of different ways. Some organizations choose to create a custom content repository that is built within a complex tool like Oxygen, the structured authoring tool we discussed in Chapter 6. Some organizations have multiple types of content governance that

Content Governance Documents

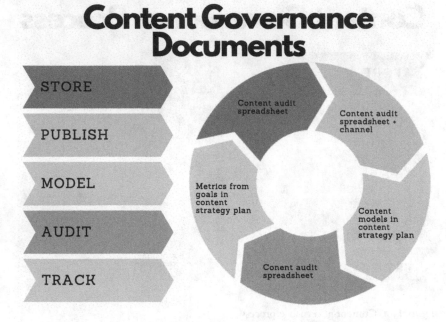

STORE

PUBLISH

MODEL

AUDIT

TRACK

Content audit
spreadsheet

Content audit
spreadsheet +
channel

Content
models in
content
strategy plan

Conent audit
spreadsheet

Metrics from
goals in
content
strategy plan

Figure 11.2 Content governance documents

are each assigned to different teams (e.g., internal documentation, marketing, customer support). Some organizations use a custom database to audit their content.

The overall process we've described is sound, however, and covers the main elements of content governance. How an individual organization *implements* content governance will strongly depend on the specific goals, tools, and audience needs of that organization.

Retooling Content for Future Use

As you have probably come to realize, creating content can be a costly effort in terms of the time it takes to get it right. This is why whenever you can reuse the quality content you've created, you want to. If you're creating content from scratch every time a new organizational goal or audience need arises, you're most likely working too hard.

You can work smarter by developing evergreen content and editing/revising older content.

While we always want to make sure our content is up to date, accurate, and effective for our purposes, there are also different ways of reusing and re-tooling a lot of our content so that we don't have to create everything from scratch every time.

Developing Evergreen Content

The first approach to content reuse is developing what is called *evergreen content*. Just like evergreen trees that stay green and fresh all year round, evergreen content is content that is always fresh for your audience and can thus be reused in new publications without serious revision.

A simple example of this is a snippet of content that describes the overall aims of a company's product. Say the company is focused on maintaining an application programming interface (API). An API is essentially an extension to an existing application that allows organizations to add new features to the application. Almost all major technology companies have them. They include:

- Twitter bots
- log-in using XYZ
- weather snippers
- pay with PayPal
- Google Maps
- travel booking
- e-commerce

(Brewster, n.d.).

So, say our organization is a company selling an API that allows companies to create interactive technical documentation for use on their websites (a tool like this: https://www.dozuki.com/).

The overall aims of this product are unlikely to change. So the goals of the product, meaning a description of *how it enables companies to create interactive technical documentation for use on their websites* is probably going to be evergreen.

This is valuable content to put in an authoritative repository. Once it's in there, it can be used throughout the organization to describe the goals of the product. It can be put on the company's website, published to the company's blog, used in internal documentation, and published to the company's social media.

The main factor as to whether content is evergreen or not is time. Time-sensitive content, such as an announcement of a rollout of new features for Dozuki, is not going to be evergreen. This one-off content will only be relevant for a specific window in time, such as the months leading up to the release and a few months after the release.

Every organization needs both kinds of content. It's not realistic to make all your content evergreen. However, *the more of your content you can make evergreen, the less revision you'll have to do later.* See how that works? Imagine if, instead of developing this content without time-sensitive information, Dozuki included information such as release dates, new features, and updates to system requirements in their description of the overall aims of the product. Now, they've turned content that could have been evergreen into content they have to update every time one of these time-sensitive pieces of information becomes outdated.

While evergreen content doesn't diminish the importance of producing regular content by any means, it does highlight the fact that there are ways that you can make your content work harder for your organization. Here are some other examples of evergreen content:

- answers to frequently asked questions from the perspective of different audience segments
- industry tips or how-to explanations, and
- explanations of common industry concepts that are important and relevant to your organization

(*The beginner's guide*)

Even evergreen content needs to be evaluated as part of content governance to ensure it stays relevant, however. And that time-sensitive content? That's the content that is most likely to become irrelevant, of course, which is why you also need to be revising and editing your older content.

Revising and Editing Older Content

Although not all of our content can reach the threshold of being evergreen, there are other strategies for reusing older content in order to give it a freshening up for our audience. This will involve revising and editing outdated content to make it current.

There are a few reasons why we would want to do this.

First, older content might be incorrect or irrelevant. Maybe it includes outdated information about what services and products our organization offers. Maybe we have webpages that have broken links, images, and other lost media. We want our audience to continue to view our content as accurate and trustworthy, so going back through our older content can help ensure that everything is as relevant as possible.

Second, revising and editing older content can help with our organization's SEO rankings. As new content gets published and rises to the top of internet searches, some of our old content might start getting pushed down the page, even if the content is still helpful and relevant. A key factor in SEO rankings is how recently the content was updated. Some content may only need a minor tweak, other content may need a total overhaul. A lot will depend on any changes happening within your organization. Either way, having content that shows it was updated recently, rather than 5+ years ago, can go a long way toward encouraging an audience to trust our content is up to date and relevant.

Last, we can consider retooling old content for a new audience or a new genre or channel. Think back to a piece of content that you wrote a few months or even years ago. When you were writing that content, you were probably thinking about your audience (like a good content strategist would), what the purpose of the content was, and what your audience might accomplish with the content. Over the course of several months or years, that content may no longer be relevant to that audience (e.g., existing customers). It may have become relevant to a different audience, however (e.g., new customers). Rather than simply retire the content, why not revise it for this new audience? Often, retooling older content for a new audience can help save you time as compared to having to develop completely new content from scratch.

Similarly, sometimes content that has outlived its usefulness on an existing channel (e.g., social media) can experience resurgence on a new channel (e.g., email). Maybe your organization developed a written how-to guide for a product. As time goes on, you notice that there are fewer and fewer views of that guide where it is currently hosted: the customer support section of your website. At the same time, however, you notice a lot of your competitors are creating video content to engage audience members on a channel like YouTube. Why not treat that guide as a script for a short 2–3-minute video?

This could kill two birds with one stone: it could draw a new audience of people to your organization while also boosting the life of the how-to guide.

Regardless of how you do it, content governance is the only way to control the content chaos that many organizations experience. Neglect it at your own peril.

Getting Started Guide: Creating a Content Governance Plan

As part of your content strategy plan, you should include the core governance activities mentioned in this chapter, most likely within your content goals and your editorial calendar, though we've seen content strategists also include these activities in individual content models. Regardless, here are some steps to get you started:

1. Store: Decide on an authoritative repository for all content. What will be the go-to source everyone in your organization uses? A database? A spreadsheet? A CMS? Some combination? As long as everyone agrees on it, the specific location isn't important.
2. Publish: Decide what process you will use when publishing new content to ensure the authoritative repository is used. Will all content need to go through a central review committee before it gets published? Will people have to log in to a specific tool that's attached to your repository in order to develop new content? Will there be one final review step before any piece of content gets

published that includes matching the content to what's in the repository?

- Model: Be sure that your content models are current and encourage effective governance. Do they mention the authoritative repository? Do they require content developers to make use of the repository? Do they include a step for documenting the new content and its location in the repository? Do they match what's in the repository with current organizational goals and audience needs?
- Audit: Add an ongoing content audit to your content strategy plan, at least yearly. For faster-moving content, such as social media posts or time-sensitive blog posts, you may need to review it more often, such as monthly or quarterly. The place this task should probably be mentioned is as a recurring event in your editorial calendar.
- Track: Track metrics to ensure you're on the right track. These metrics should be included in your content goals and should be easily accessible through organization reports, analytics programs, and other sources of data. Again, there's no hard-and-fast rule for how often you should assess your metrics. Many organizations choose to do so monthly, if not quarterly.

References

Brewster, C. (n.d.). *7 examples of apis in use today*. Trio. Retrieved April 3, 2022, from https://trio.dev/blog/api-examples

The beginner's guide to evergreen content. Digital Marketing Institute. (2021, July 2). Retrieved April 3, 2022, from https://digitalmarketinginstitute.com/blog/the-beginners-guide-to-evergreen-content

Further Reading

Baldwin, E. (2021, December 23). *Content governance: Principles for before and after creating content*. GatherContent. Retrieved April 3, 2022, from https://gathercontent.com/blog/content-governance-people

Rago, M. (2021, July 27). *Content governance: What it `is and how to get started*. 18F. Retrieved April 3, 2022, from https://18f.gsa.gov/2021/07/27/content_governance_what_it_is_and_how_to_get_started/

Speiser, M. (2020, July 13). *The essentials of content governance*. Knotch: Pros & Content. Retrieved April 3, 2022, from https://prosandcontent.knotch.com/posts/content-governance

Welchman, L. (2015). *Managing chaos: Digital governance by design*. Brooklyn, NY: Rosenfeld Media.

12 Localizing Content

Many years ago, one of the authors of this book worked in tandem with several other Americans with a non-profit organization in Chiclayo, Peru. Although each of us knew a little bit of conversational Spanish, we sought the services of a translator to ensure effective and clear communication for larger presentations within the city. During one of our large presentations, one of the Americans decided to tell a joke to the audience through the translator.

Although nearly all ideas and concepts are translatable between languages, it is usually an unspoken rule not to tell jokes in a live translation setting because jokes do not translate well into another language and culture. This is because languages don't always have a word-for-word translation. Also, culture is a complex group of ideas that includes history, social norms, and a specific geographic context. Because of these social and cultural differences between languages in a live translation setting, it is extremely difficult to convey a joke from one language into another language.

The speaker ignored all this and decided that this joke simply had to be told in the middle of the presentation. When the translator reached the end of the punchline, the audience exploded into laughter, far more than the author of this book would have anticipated, even from an English-speaking audience, let alone through a translation. After the presentation, the translator explained that instead of directly translating the joke, which would not be effective or humorous to the audience due to differences in culture and language, they told the audience that the speaker was telling a hilarious joke and to laugh accordingly.

We tell this story not to mock the well-meaning presenter, but to emphasize that organizations can't simply decide to take their existing content from one cultural context and then distribute it to another cultural context without a lot of effort. At the same time, many organizations are attempting to do just that: to repurpose their content into a new cultural context. Many large corporations now reach global audiences. Even non-profits sometimes have sister organizations that span multiple countries.

So, content *does* sometimes have to move between cultural contexts. It does have to reach a global audience. This process is called *localization*.

DOI: 10.4324/9781003164807-12

To successfully localize content, the differences in cultures must be taken into consideration in addition to translating content from one language to another. What is needed is a global content strategy.

What Is a Global Content Strategy?

According to Swisher (2014), "[a] global content strategy is a plan for managing content that is intended for people whose main language is something other than the source language" (p. 2). This means that a new plan and new content will need to be created for each new audience. In the previous example, to effectively reach the audience in Chiclayo, a modified version of the American joke would need to be created in Spanish for the people of that city. If the non-profit wanted to expand presentations to Brazil, a modified joke in Portuguese would be needed to effectively reach that audience. To take it a step further, the non-profit may also decide to create a presentation script in both Spanish and Portuguese to make sure the presentations are consistent, and that no on-the-spot translation decisions have to be made such as in the previous example.

Why Is Having a Plan for Global Content Important?

Having a plan in global content strategy plan in place facilitates the creation of consistent content throughout different audiences, countries, and languages. In many cases, the global content needs of organizations are significantly more complex than a single presentation. For example, a company's global website has on average 28 different languages (Swisher, 2014, p. v). This means that a company that currently only has content in their source language would need to expand their content by 28 times on average in order to reach a global audience. Think about all the elements of a content strategy plan described in Chapter 7, and you see the problem. Multiplying the goals, audiences, channels, content models, and editorial calendars of an organization by 28 equals a lot of planning.

At the same time, not having a plan would mean creating 28 times the content with no guidelines as to how to do so. Creating a global content strategy plan at the onset of this process will help save time, money, and frustrations later.

Conducting a Global Content Audit

A global content audit is very similar to the process we described in Chapter 5, but it also includes content in different languages, who is creating content, who is reviewing content, translations vendors (if applicable), and TMs (translation memories, which we discuss later). Another key consideration when creating a global content strategy plan, however, is the status of your organization's content repository. Multiplying the amount of content means multiplying the amount of space required to store that content. If the content is tied to a specific tool, such as a content management system (CMS), then there is also the question of whether the CMS will allow for easy translation or whether it is designed to center on the culture of its origin.

If your content will be housed in something like a web-based CMS that will provide the same content in a variety of languages, then the first step of a global content audit is to determine how easy your audience can find the language they need. In general, the goal is for your audience to find the language they need in as little time as possible (Swisher, 2014, p. 59). This is normally accomplished via a global splash gateway, where the homepage displays a list of regions, countries, and languages to select, or a global gateway button on the homepage. This button features a dropdown list of regions, countries, and native languages.

The Process of Localizing Content

As mentioned before, localization is *the process of adapting content from a source culture and language into a different culture and language.* The key to localization is recruiting native speakers for a specific culture and language. In most cases, organizations need content developers that are native to the culture and have extensive knowledge of the language. It is the job of the localization team to make sure new content is appropriate to the culture and language of a specific audience.

Since direct translations are not always possible between languages, some organizations give their localization team too much leeway to operate independently of the source language, such as by removing pieces of content that simply aren't intelligible within a specific cultural context. This could result in inconsistent content between the source language and the localized language, which could lead to frustration from audience members within the localized culture. One way to address this is to involve different regions and localization teams in the planning stages of each content strategy project. Though it might be easier to allow the localization team to create their own content in line with the project instead of trying to directly translate the source content for their region, someone needs to be in charge of checking localized content against the source content to ensure consistency of messaging.

Translation

As mentioned earlier, one-to-one translation, even when conducted by bilingual people, is often not effective because it doesn't carry the same cultural meanings from one language to another. Similarly, web translators or other machine translation tools are not accurate enough to use for business purposes. These types of translators are fast, but they do not take into consideration cultural meanings and differences between cultures. As a result, the text of a source language may be difficult to read or decipher after a machine translation into another language.

For these reasons, a localization team and/or translation vendor should be used to make sure the words and meaning of the source language effectively cross the boundaries to a different culture and language.

One way to save time, money, and frustrations with the translation process is to edit the source language before it is submitted for translation. This includes:

- shortening sentences
- eliminating unneeded words (using concise wording)
- using consistent terminology
- eliminating idioms, jargon, and slang
- ensuring grammar is correct, according to the rules of the source language

Additionally, if you are using multiple translation vendors, it is important to maintain ownership of what is called a *translation memory*. A translation memory is "a database that stores your completed and approved translations so that the translations can be reused for future projects" (Venga, as cited in Swisher, 2014, p. 56). This is helpful because it helps an organization track translations they've already approved for a specific language and culture so that these translations don't have to be redone for another project.

Also, an approved and consistent translation used throughout multiple projects ensures that certain words and phrases will be translated consistently throughout the content lifecycle, as opposed to a translator creating multiple translations for a word or phrase that mean generally the same thing but aren't an exact match. For your translation memories to be effective, however, they must be updated frequently to ensure they are consistent with what the translation vendor is receiving from your organization.

Transcreation

Through the use of translation and a global content strategy plan, a large amount of content can be reused and repurposed to multiple regions, countries, and languages. However, some content can't be as easily reused and new content may need to be created to reach each specific audience. This is where transcreation comes in. "To create a global marketing campaign that evokes the desired response in every culture you target, you need to recreate the campaign—the words and images—for every culture" (Swisher, 2014, p. 21). While it's important to make content as simple, and thus as universal, as possible, in other words, oversimplified content doesn't really engage an audience as effectively as content tailored for that audience. The style of content needs to be adapted to specific cultural contexts in order to successfully connect with audience members within that context.

For example, images and videos are not always universal and, in many cases, can't be reused between cultures due to differences in the ways different cultures relate to these forms of content. Additionally, symbols, gestures, and sounds do not always translate across cultures. To help ease the process of transcreation, these items in the source language should be tested for universality and reviewed by your different localization teams. This will allow for more content to be used between audiences and will thus create less of a need for content to be created separately.

Similar to translation, transcreation becomes significantly easier if the source language is as clear and concise as possible. This ensures that the developers of new content in a new language aren't bogged down by ambiguity.

Getting Started Guide: Developing a Global Content Strategy Plan

So, obviously developing a global content strategy plan is a complex process. This is why there are entire content strategy companies that specialize in this form of planning (i.e., https://contentrules.com/). There's a lot to manage with a global content strategy plan, but you can get started on one by reviewing the information in this chapter and following these steps:

1. The first step in developing a global content strategy plan is to define who your audiences are that are outside your current cultural context. You'll want to do audience analysis, similar to what we described in Chapter 3, but for global audiences. This probably means you'll need to start building a localization team from the get-go, unless you already have contact with members of the target cultures.
2. From here, you need to proceed with the content strategy planning process described in Chapter 7, but with a focus on responding to the goals, audiences, channels, content models, and editorial calendar of your target cultures. The same channel won't work the same way in every culture. You have to assume that no part of your content strategy will be universal, in fact, which is why engaging a localization expert familiar with the target culture is so important throughout this process.
3. As we described in this chapter, you then need to develop a separate content strategy plan for each target culture you plan to engage. These plans should work in tandem with your main content strategy plan, of course, but should also be responsive to the specific requirements of each target culture and language.

4. Once you have a content strategy plan for each target culture, it's time to decide what content can be translated and what content needs to be transcreated. Once you segment your content into these two groups, you'll want to engage a translation vendor to start translating, and your localization expert to start transcreating.
5. Don't forget the usability, accessibility, and governance guidelines we recommended in Chapters 10 and 11! All audiences want useful, accessible content. And content can become outdated, no matter the language or culture.

References

Swisher, V. (2014). *Global content strategy: A primer*. XML Press.

Further Reading

Dutta, S. (2020, September 8). *The 5 do's and don'ts of global content creation and transcreation*. Skyword. Retrieved April 3, 2022, from www.skyword.com/contentstandard/the-5-dos-and-donts-of-global-content-creation-and-transcreation/

Evans, N. (2018, May 18). *Effective global content strategy—it's more than translation*. CMS Connected. Retrieved April 3, 2022, from www.cms-connected.com/News-Archive/May-2018/Effective-Global-Content-Strategy-It%E2%80%99s-More-Than-Translation

How to create a global content strategy. Memsource website. (n.d.). Retrieved April 3, 2022, from www.memsource.com/blog/how-to-create-a-global-content-strategy/

13 Content Tools and Technologies

Similar to your first visit to a large hardware store, this chapter may seem a little intimidating at first. You will notice a lot of different types of tools, with the names of specific tools listed under each type. The purpose of this chapter is to help you gain a general understanding of the types of tools used by content strategists, to be able to discern which specific tool is best for particular workplace situations, to learn how technology and communication practices shape one another, and to understand the importance of being able to learn a new tool effectively.

> We're going to stick with the term tool throughout this chapter, by which we simply mean *a technology applied for a specific purpose.*

Going back to the analogy of visiting a large hardware store, you may have heard of some of the tools listed before, and most likely, there will be tools you've used to create, plan, manage, or analyze content that aren't mentioned. Our goal in this chapter is *not* to create a comprehensive list of all the tools a content strategist might ever use. Rather, it's to encourage you to build your own toolkit for content strategy by introducing different types of tools common to this field.

The challenge with a chapter like this is that tools shift and change constantly. Chances are, some of the tools we mention in this chapter will be outdated by the time this book reaches you. And that's perfectly fine. Most content strategists, and really most industry professionals, start their careers with just a few tools in their toolkit. As situations call for it, content strategists will learn additional tools, and as time goes on, they will become effective with a large variety of tools that cover many different types of situations. Since new tools are constantly released, maintaining your content strategy toolkit is an ongoing learning process that is tied to the specific problems you end up solving for organizations.

Before we discuss the different types of tools used by content strategists, we would like to discuss the guiding framework we will be using for this

DOI: 10.4324/9781003164807-13

chapter. We will be using the technological literacy framework of Hovde & Renguette (2017) to explore different tools. The framework can be summarized by four main concepts:

- functional literacy—being able to use a specific technology effectively
- conceptual literacy—knowing general concepts of technology type
- evaluative literacy—choosing the best technology for a specific workplace situation
- critical literacy—understanding how technology and communication practices shape one another

Functional Literacy and Content Strategy Tools

In order for a content strategist to be able to complete all the tasks we've described throughout this book, whether those are tasks associated with audience analysis, content auditing, content strategy planning, content development, or content editing, they'll need to use various tools, from website analytics programs to desktop publishing software. Sometimes this will require content strategists to do quick, on-the-spot learning in order to understand how a given tool functions. Prepare for that. Don't expect to be good at every new tool instantaneously. That's what online tutorials are for.

Conceptual Literacy and Content Strategy Tools

As you will see later in the chapter, there are a lot of different types of tools that content strategists use. While it is possible to develop a functional literacy of any tool that you use regularly, it is impossible to be equally versed in every single tool on the market, not to mention future updates of new features that are rolled out within familiar tools you use daily. Because of this, it is useful to have conceptual literacy of different technology types, which includes understanding the roles of different types of tools in the content strategy process. That's why we contextualize each category of tool here so that you can develop a working knowledge of how each one functions in the greater content strategy landscape.

Evaluative Literacy and Content Strategy Tools

Developing conceptual literacy means gaining knowledge that helps you categorize the role and function of different tools. And building off of that, evaluative literacy enables you to choose the best tool for a specific workplace situation. If you have learned nothing else from this book: content strategists are jacks and jills of all trades. They have to manage a dizzying array of skill sets, knowledge types, and tool types. If your experience of working in content strategy is anything like ours, you will spend a significant amount of time researching tools that will make your work easier and more efficient, which include comparing different types of tools to one another. This evaluation process is simply part of any technology-driven field.

Critical Literacy and Content Strategy Tools

Finally, developing critical literacy means being able to understand the context in which a tool is best used, to understand how to use tools in different contexts, and to understand how to learn new tools quickly when required. Like critical thinking, this form of literacy takes a lot of higher-order analysis. This analysis should be guided by questions like:

- Should you use Tool Type A (e.g., a desktop publishing tool) or Tool Type B (e.g., an authoring tool)?
- What's the difference between tools?
- What does each tool allow you to do and prevent you from doing?
- What's the learning curve like for each tool?
- What context does each tool best fit into?
- Which tool is most appropriate, given the context you're currently working in (e.g., a small non-profit versus a major corporation versus a university)?

This is the hardest form of literacy to acquire, and the one that people who are new to a technical field struggle with the most. You should also expect that: struggle. Learning is hard, whether you're trying to use an API for the first time or to build your first persona.

And the best teacher, in our opinion, is *experience*. Don't try to tackle every content-related problem in existence. Start with the simplest one first and work from there. And ask your colleagues what they've done in similar situations. Ask your fellow content strategists.

You'll be surprised how quickly you become extremely literate in all the tools at your disposal. Next we give you a jump start on this process by going through some of the most common tools content strategists use.

Categories of Content Strategy Tools

Desktop Publishing

Desktop publishing tools include any software that allows the user to create professional-grade deliverables with texts and graphics. Most desktop publishing tools are accessible to a range of users by providing templates for novices, as well as advanced editing options for expert users. Some of the most common desktop publishing tools can be divided into three types: word processors, spreadsheets, and presentations.

Word Processors

- Microsoft Word (https://www.microsoft.com/en-us/microsoft-365/word)
- Google Docs (https://www.google.com/docs/about/)
- Apache OpenOffice Writer (https://www.openoffice.org/product/writer.html)

Spreadsheets

- Microsoft Excel (https://www.microsoft.com/en-us/microsoft-365/excel)
- Google Sheets (https://www.google.com/sheets/about/)
- Apache OpenOffice Calc (https://www.openoffice.org/product/calc.html)

Presentations

- Microsoft PowerPoint (https://www.microsoft.com/en-us/microsoft-365/powerpoint)
- Google Slides (https://www.google.com/slides/about/)
- Apache OpenOffice Impress (https://www.openoffice.org/product/impress.html)
- Prezi (https://prezi.com/)

Ultimately, the type of desktop publishing tool you use will be determined by the preferences and available resources of your organization—that is, your workplace or school. For example, if you work for a smaller organization that needs to be budget-conscious, it is likely that you will be using a publishing tool that is freely available like OpenOffice. Many larger organizations buy Microsoft licenses because they get a deal when they buy for the entire organization. Sometimes individuals prefer certain tools over others. As long as the tool meets the requirements of your current task, and is a permitted tool within your organization, it doesn't matter where it comes from or who built it.

Collaboration Tools

Though work-from-home has been a movement for many in the technology sector, the COVID19 pandemic has shifted this trend into high gear. Since people working from home still need to be able to collaborate on projects, tools that enable virtual meetings, calls, and other types of interaction have soared in popularity.

For the sake of this chapter, we will be considering a collaboration tool to be "any piece of software that helps people get work done together. They let people know about team activity on work that pertains to them" (Duffy, 2022). Because there are a wide range of collaboration tools, we will be exploring them through the following categories: collaborative writing tools, visual collaboration and brainstorming tools, videoconferencing, team messaging, appointment and meeting scheduling, and project management.

Before we explore specific tools, however, it's important to note that collaboration tools, in particular, should be chosen based on an organization's needs, budget, and culture.

> Throwing a new tool at a bunch of people and telling them to use it instead of email doesn't work. To start using a collaboration tool successfully, all the key players on the team need to buy into it. It has to become part of the culture.
>
> (Duffy, 2022)

Therefore, it is important to understand how an organization collaborates and communicates in order to find a collaboration tool that will fit into those existing processes. If the existing processes work effectively, there really isn't a need to overhaul them. Any substantial change to an organization's communication style will slow down productivity, at least temporarily, because it will take time for individuals within the organization to accept and adapt to the changes. The collaboration tool should fit into the organization's culture, in other words, instead of the culture having to completely change to accommodate the tool.

Collaborative writing technologies

- Microsoft OneNote (https://www.microsoft.com/en-us/microsoft-365/onenote/digital-note-taking-app)
- Google Docs (https://www.google.com/docs/about/)

Visual collaboration and brainstorming tools

- Miro (https://miro.com/app/dashboard/)
- Conceptboard (https://conceptboard.com/)
- Lucidspark (https://lucidspark.com/)

Videoconferencing

- Zoom (https://zoom.us/)
- Google Meet (https://meet.google.com/)
- GoTo Meeting (https://www.goto.com/meeting/join)

Team messaging

- Slack (https://slack.com/)
- Microsoft Teams (https://www.microsoft.com/en-us/microsoft-teams/group-chat-software)
- Google Chat (https://workspace.google.com/products/chat/)

Appointment and meeting scheduling

- Appointlet: https://www.appointlet.com/
- Calendly: https://calendly.com/
- Doodle https://doodle.com/
- When2meet: https://www.when2meet.com/

Project management software

- Liquid Planner (https://www.liquidplanner.com/)
- Teamwork (https://www.teamwork.com/)
- Zoho Projects (https://www.zoho.com/projects/)
- Trello (https://trello.com/)
- Jira (https://www.atlassian.com/software/jira)

Authoring Tools

Authoring tools are very similar to desktop publishing software in that both allow the user to create deliverables that can be saved in a variety of formats. However, that is where the similarities end. While desktop publishing software has a very low learning curve for anyone familiar with basic computer technology, authoring tools are specialized to fit the needs of technical communicators, and have a notoriously high learning curve.

With this learning curve comes the ability to manage content at a scale that is simply impossible (or would be very, very time-consuming) with desktop publishing software. Authoring tools are used by people managing large amounts of technical content across multiple channels. This means they can be very useful, however, if your role as a content strategist entails this kind of work.

They largely fall within four categories: structured authoring, help authoring, API and developer documentation, and all-in-ones (Mayr, 2018). As these categories can be confusing for people who don't deal with technical content on a regular basis, we provide brief summaries for each category here with our lists of actual tools.

Structured Authoring

Structured authoring is that process we've discussed a few times throughout this book (and most extensively in Chapter 6) whereby a content developer can create structured content, meaning content broken down to its most basic components, that can be reused on demand and published in a variety of formats without having to worry about formatting. These tools are best used by content developers working with large amounts of technical content that needs to be reused within a lot of different content genres.

- Adobe Framemaker (https://www.adobe.com/products/framemaker.html)
- Oxygen (https://www.oxygenxml.com/)

- XMetaL (https://xmetal.com/)
- Arbortest (https://www.ptc.com/en/products/arbortext)

Help Authoring

Like the name suggests, these tools help content developers who want to create help documentation for a product or service. Look at your latest help forum for a large company like Google or IBM and you'll see why: there are thousands upon thousands of pages of content in those forums. Some is user-generated, some is maintained by the organization itself, some is moderated by power users of the software. The point is, maintaining a large database of help documentation is extremely difficult, time-consuming work. These tools make this job a lot easier.

- Madcap Flare (https://www.madcapsoftware.com/products/flare/)
- Adobe RoboHelp (https://www.adobe.com/products/robohelp.html)
- Help+Manual (https://www.helpandmanual.com/)
- ClickHelp (https://clickhelp.com/)
- Madcap Doc-to-Help (https://www.doctohelp.com/)
- Dozuki (https://www.dozuki.com/)

API and Developer Documentation

This category of tools applies to a trend in content called *docs as code* where technical content creators use the same code an application runs on to document how to use the application. This ensures that as new features are released content creators adapt their documentation. It's kind of like these content creators are building their own version of the application, but the point is to explain what the application does, rather than to actually run the application itself. This is obviously a very technical area of content creation that requires significant expertise.

- Sphinx (https://www.sphinx-doc.org/en/master/)
- Swagger (https://swagger.io/)
- Jekyll (https://jekyllrb.com/)
- Slate (https://github.com/slatedocs/slate)

All-in-Ones

These large tools come with a lot of features (essentially all the ones listed in the authoring tools section) and large price tags to boot. They are typically used by very large organizations with very complex content needs. They can control everything from an organization's internal content repository to its localization efforts to its customer-facing documentation to its website. Many of these tools function like massive content manage systems (CMSs), but they come bundled with many more features than something like WordPress, so they are better classified as Component-based Content Management Systems (CCMSs; see later) because they break down content into such granular detail and allow for such fine-grained control of it. To get a sense of just one of these tools and what it can do for an organization, we recommend reading O'Keefe (2017) and her explanation of Adobe Experience Manager's features.

- Author-it (https://www.author-it.com/)
- Heretto (https://heretto.com/)
- Tridion Docs (https://www.rws.com/content-management/tridion/docs/)
- IXIASOFT (https://www.ixiasoft.com/)
- Adobe Experience Manager (https://business.adobe.com/products/experience-manager/adobe-experience-manager.html)

Content Management Systems (CMSs)

A content management system is "software that helps create, organize, and maintain digital content" (Jones, n.d.). This information is usually stored in the form of pages, posts, tags, and categories for future use. Many websites are now built using CMSs because they save organizations significant development time over building a website from scratch. And many CMSs are also open source, meaning new features can be added to them as they make their source code available (see next).

- WordPress (https://wordpress.org/)
- Drupal (https://www.drupal.org/)
- Joomla! (https://www.joomla.org/)

Component-Based Content Management Systems (CCMSs)

Component-based content management systems (CCMSs) are a type of CMS that specializes in organizing content by content type as well as allowing advanced publishing options and roles for content developers (e.g., author, reviewer, manager, IT specialist). Several of the all-in-one tools we included here (Heretto, Tridion, and IXIASOFT) are best classified as CCMSs by the way they enable users to manage content.

The Role of Open Source

You may have heard some tools referred to as *open source*. This means that their source code is available for redevelopment. If you purchase a license for Adobe Experience Manager, just as one example, you are purchasing a license to a *proprietary* tool, meaning you are essentially renting that tool for usage. When you purchase a license for an open-source tool, however, you gain access to the source code for the technology, meaning that you essentially own the full tool and can even redevelop it into something else, barring some exceptions (*Proprietary*, 2019). There are three main open-source tools that are used in content strategy: markup languages and stylesheet languages, information architectures, and software.

Markup Languages and Stylesheet Languages

Markup languages are a form of code that structure how digital documents display content. Any time you load a website, you are relying on a markup language, most likely HTML. For content strategists, markup languages do lots of things, from helping to structure website content to enabling them to create internal or external documentation. Stylesheets do the opposite of markup languages: they control the *presentation* or *design* of digital content (think color, layout, device-specific displays such as mobile, etc.). Don't get us wrong: content strategists aren't web developers. However, many content strategists deal with code in a more superficial fashion than developers do as they work to manage digital content.

- HTML (https://developer.mozilla.org/en-US/docs/Learn/Getting_started_with_the_web/HTML_basics)
- XML (https://developer.mozilla.org/en-US/docs/Web/XML/XML_introduction)
- CSS (https://developer.mozilla.org/en-US/docs/Web/CSS)

Information Architectures

Open-source information architectures help organizations structure content in such a way that it can be used by a variety of other people and technologies, including different types of professionals (e.g., technical writers, content strategists, UX designers) and different types of tools (e.g., desktop publishing, authoring tools, CMSs). Several of the other tools we've mentioned thus far run on open-source information architectures (i.e., Oxygen, Dozuki, Framemaker, XMetaL, Madcap Flare, Heretto, and IXIASOFT).

- DITA XML (http://dita.xml.org/)
- oManual (https://www.omanual.org/)

Software

Open-source software programs are full-blown applications that provide a variety of features out of the box. Many of the tools we've mentioned thus far are open source, including all of the CMSs, OpenOffice, and Dozuki.

Application Programming Interfaces (APIs)

Application programming interfaces (APIs) are tools within existing applications that enable writers to build their own, simpler applications, often for the purposes of storing content within those larger applications. We've heard of many organizations who use open-source tools as a basis to solve complex content problems, such as using WordPress APIs to create a custom CMS for storing internal documentation.

As we mentioned in Chapter 11, APIs are everywhere and include:

- Twitter bots
- log-in using XYZ
- weather snippers
- pay with PayPal
- Google Maps
- travel booking
- e-commerce

(Brewster, n.d.).

Multimedia Content Creation Tools

Many content strategists deal with multimedia content, but few of us are full-on graphic designers. We need tools that help us quickly and easily develop high-quality videos, audio files, photos, images, and document designs. Some of these tools are easier to use than others, and some are free while others come with a significant price tag. We've classified them here by the type of media they produce.

Video

- Adobe Premiere (https://www.adobe.com/products/premiere.html)
- Camtasia (https://www.techsmith.com/video-editor.html)

Audio

- Adobe Audition (https://www.adobe.com/products/audition.html)
- Audacity (https://www.audacityteam.org/)

Photo

- Pixabay (https://pixabay.com/)
- Shutterstock (https://www.shutterstock.com/)
- Adobe Photoshop (https://www.adobe.com/products/photoshop.html)
- Pixlr (https://pixlr.com/)
- GIMP (https://www.gimp.org/)

Image

- Adobe Photoshop
- Pixlr
- GIMP
- Adobe Illustrator (https://www.adobe.com/products/illustrator.html)
- Canva (https://www.canva.com/)

Document design

- Canva
- Microsoft Publisher (https://www.microsoft.com/en-us/microsoft-365/publisher)
- Adobe InDesign (https://www.adobe.com/products/indesign.html)

Channel-Specific Tools

Many of the content channels we've mentioned throughout this book, including websites, search engines, social media, and email, have grown so complex that tools have been developed just to manage them. These tools shouldn't replace an authoritative content repository, but they are encouraged to help content strategists who are publishing content to specific channels on a regular basis.

Websites

- Any of the CMSs or CCMSs we mentioned earlier are recommended for controlling web content. From a content perspective, we don't recommend that most organizations develop a fully custom website from scratch, simply because it makes content very difficult to manage. The exception to this is very large organizations with equally large budgets who can develop their own custom CMS right along with their website.

SEO

- Ahrefs (https://ahrefs.com7)
- Google Search Console (https://search.google.com/search-console/about)
- SEMRush (https://www.semrush.com/)
- KWFinder (https://kwfinder.com/)
- Moz (https://moz.com/)

Social media

- Hootsuite (https://www.hootsuite.com/)
- Buffer (https://buffer.com/)

- MeetEdgar (https://meetedgar.com/)
- SocialPilot (https://www.socialpilot.co/)
- Sendible (https://www.sendible.com/)

Email

- MailChimp (https://mailchimp.com/)
- ActiveCampaign (https://www.activecampaign.com/)
- MailerLite (https://www.mailerlite.com/)
- Moosend (https://moosend.com/)
- Drip (https://www.drip.com/)

Analytics Tools

As we mentioned in Chapter 3, analytics tools are a great way to measure the impact of content on an audience. They also tend to be channel-specific, but we included them as a separate category because they really function differently than a content management tool, even though you'll see several of the tools listed here repeated because they include analytic features:

- Google Analytics (https://analytics.google.com/analytics/web/)
- Facebook Audience Insights (https://www.facebook.com/business/insights/tools/audience-insights)
- Twitter Analytics (https://analytics.twitter.com/about)
- LinkedIn Analytics (https://www.linkedin.com/help/linkedin/answer/a547077/linkedin-page-analytics-overview)
- Hootsuite Analyze (https://www.hootsuite.com/platform/analyze)
- Mailchimp Analyze (https://mailchimp.com/features/reports-and-analytics/)

Miscellaneous Tools

We've also included a category for miscellaneous tools that we've personally found to be helpful (and that other content strategists we've spoken to have as well). These range from tools for creating personas to tools for managing citations.

Persona Creation Tools

Remember those personas we mentioned in Chapter 3? You can certainly design them on your own using one of the tools in this chapter, *or* you can use one of the many free online persona templates.

- Xtensio User Persona Template (https://xtensio.com/user-persona-template/)
- Hubspot's Make My Persona (https://www.hubspot.com/make-my-persona)

Grammar Checker

Most word processing tools (and many authoring tools) include built-in spell checkers, but what about that Facebook post you're working on? What about that email? Grammarly checks all content that exists within a web browser, making it an invaluable tool for any content strategist.

- Grammarly (https://www.grammarly.com/)

Citation Manager

If you're planning to be (or are) a content strategist in an industry that uses a very specific style guide, such as Chicago or APA, consider using Citation Machine. It's not perfect, but it will save you a lot of time, especially if you're juggling multiple citation styles.

- Citation Machine (https://www.citationmachine.net/)

Getting Started Guide: Building Your First Content Strategy Toolkit

We know this chapter may be daunting, but just remember what we said earlier: tools are supposed to make your life easier, not harder. If you have a content-related problem, find a tool to help you solve it. If you can't find one or one doesn't fit your needs, use something simple that you're familiar with.

You can start building your technological literacy within the field of content strategy by brainstorming the toolkit you might use right now. The following questions will help you:

- Functional: What are you trying to do, specifically, with content?
- Conceptual: What categories of tools from this chapter seem to fit with what you are trying to do?
- Evaluative: Which of the tools from each category seem like the best fit for your needs and those of your organization?
- Critical: What do you need to learn about each tool in your toolkit to put into use? What kind of workflow can you imagine that will help you make use of the tools?

References

Brewster, C. (n.d.). *7 examples of apis in use today*. Trio. Retrieved April 3, 2022, from https://trio.dev/blog/api-examples

Duffy, J. (2022, January 4). *The best online collaboration software for 2022*. PCMag. www.pcmag.com/picks/the-best-online-collaboration-software?test_uuid=001OQhoHLB xsrrrMgWU3gQF&test_variant=a

Hovde, M. & Renguette, C. (2017). Technological literacy: A framework for teaching technical communication software tools. *Technical Communication Quarterly*, 26(4), 395–411.

Jones. S. (n.d.). *5 types of content management systems (CMS)*. Ixiasoft. www.ixiasoft.com/types-of-content-management-systems/

Mayr, C. (2018, November 21). *The 5 authoring tools technical writers use*. STC Carolina. www.stc-carolina.org/2018/11/21/the-5-types-of-authoring-tools-technical-writers-use/

O'Keefe, S. (2017, May). *The age of accountability: Unifying marketing and technical content with Adobe Experience Manager*. Scriptorium. www.scriptorium.com/2017/05/the-age-of-accountability-unifying-marketing-and-technical-content-with-adobe-experience-manager/

Proprietary vs. open source. OpenLogic by Perforce. (2019, August 29). Retrieved April 3, 2022, from www.openlogic.com/blog/proprietary-vs-open-source

14 Establishing Yourself as a Content Strategist

If you're reading this, then you've either made it all the way through this book or you skipped right to the end. Either way, you're interested in getting started as a content strategist, professionally. You have likely heard the buzz around this field and how much job growth it's experiencing. You may have even applied for a few jobs in the field. And, if you're like many early career content strategists we've spoken to, you may feel daunted by the wide variety of skills required by job ads you've seen.

We'll start this final chapter of the book with advice we've given to every early career job seeker in a fairly new profession: employers want a unicorn, but they'll probably end up with a workhorse. To decode this, a "unicorn" in the technology sector of the economy, where lots of jobs in fields like UX, technical communication, and content strategy exist, refers to *someone with every possible skill set*. These are people that are so rare they might as well be fictional, in other words.

When employers write a job description for a field like content strategy, they might have very little knowledge of the actual field. Sometimes they have so little knowledge that they hire a recruiting firm to write the job description for them. Regardless, if you search for "content strategist" in a job search engine like Indeed, it's likely that the vast majority of ads that pop up were written by someone who has never worked as a content strategist.

Sometimes job ads are also written with a see-what-we'll-get approach. Employers ask for the moon and see what the pool of candidates they end up with looks like. They might get lucky and get someone with 10 years of experience and every skill set in a field that many people have never even heard of when they started their careers. Most likely, however, they won't.

So, our first piece of advice when seeking to professionalize in content strategy is: don't be intimidated by job ads. Apply! Unless the job ad is asking for skill sets you don't possess at all: apply.

But while you're applying, there are some other things you can be doing to build your resume in content strategy. We turn to those things now.

DOI: 10.4324/9781003164807-14

Network, Network, Network

A survey a few years ago found that 85% of jobs are filled through networking (Adler, 2016). In our personal experience working with scores of job seekers in fields like UX, technical communication, and content strategy, this trend continues to hold. The reason is simple: in complex fields like these, it can be daunting for hiring managers to bring on new employees, not knowing if they're up to the challenge. These are not fields where skill sets like coding, designing schematics, or synthesizing materials rule. These are fields where the top skill sets are practices such as writing, audience analysis, editing, and usability testing.

They are skill sets that many people don't fully understand and are strongly qualitative and interpretive, in other words. When you go to hire a software engineer, you can look at previous software they've built or ask them to do some coding to test their skills. How do you test for the ability to do a content audit? Or the ability to analyze a group of people to develop personas?

Many UX designers, technical communicators, and content strategists get hired by people they already know because those hiring managers *trust* them. They know them, know their work, and know they have the ability to solve the complex problems they need them to solve. Most importantly, they know that the applicant in question *understands the process*. Every one of these fields we're discussing, including content strategy, has a process to it. And like the UX process or the process of creating effective technical documentation, the content strategy process is not something you can adequately depict in a resume. It's too big and complex.

Hiring managers want to know that when they hire someone to be a content strategist they can *do the content strategy process*. It's a lot of responsibility. There are a lot of moving parts. The worst thing that can happen is that a hiring manager hires someone who looks great on paper, but it turns out they can't perform the essential activities of content strategy, and then they have to let that person go and search again.

From talking to scores of hiring managers over the years, we've learned the struggles they face because of these issues. And we've learned that these managers also get really good at spotting people who don't have in-depth knowledge of a field. This is why they'll hire someone with direct experience in a field over someone with great credentials who doesn't have this experience. This is why credentials are great in content strategy but will always be trumped by experience.

Besides experience, having connections to people in the field is *essential*. We recommend your networking should take three forms: virtual, professional, and face-to-face. First, you should be skilling up in content strategy and adding your skill sets to your LinkedIn profile. LinkedIn is one of the number one ways that technical recruiters find qualified candidates in fields like content strategy. Be sure to ask people you know, including colleagues and professors, to endorse you for skill sets, because that's how the LinkedIn search engine works. Finally, connect with people on LinkedIn who are in jobs related to what you want to do. Want to be a content strategist at Google? Find content strategists at Google on LinkedIn and ask to connect. Don't be intimidated. You'll be surprised who will accept your connection request on LinkedIn.

Also, be sure to join the social media groups we recommended in Chapter 1:

Content strategists. Facebook. (n.d.). Retrieved January 20, 2022, from https://www.facebook.com/groups/132535916799137

Content strategy. LinkedIn. (n.d.). Retrieved January 20, 2022, from https://www.linkedin.com/groups/1879338/

Welcome to Content Strategy. (n.d.). Retrieved January 20, 2022, from https://community.content-strategy.com/

These groups are where a lot of content strategists meet and discuss current trends, including who's hiring. Also, don't be afraid to ask people you come across online for informational interviews. Create a free Zoom account and ask people who currently have the type of job you want to land for a 15-minute interview about how they got established. *Don't ask them to hire you.* This is a fact-finding mission to learn how they got where they are. Ask them how they broke into content strategy, what they like best about content strategy, and what is most challenging.

You'll be surprised how many people will agree.

Next, join any professional societies that contain content strategists. There isn't an overarching content strategy organization right now, unfortunately, but your best bet for a professional society is definitely the Society for Technical Communication, which we also mentioned in Chapter 1. Besides the STC's annual Tech Comm Summit, there are also two conferences devoted exclusively to content strategy:

Confab: The content strategy conference. (n.d.). Retrieved February 17, 2022, from https://www.confabevents.com/

Society for Technical Communication. (n.d.). Retrieved February 17, 2022, from https://www.stc.org/

Technical Communication Summit. (2021, December 6). Retrieved February 17, 2022, from https://summit.stc.org/

The LavaCon Content Strategy Conference. (n.d.). Retrieved February 17, 2022, from https://lavacon.org/

The conferences are pricey because they are geared toward people working in higher education and the corporate world whose organizations will pay for them to attend (though higher education, as in all things, pays far less than the corporate world . . . sigh). Get involved. Contact the organizers. Explain that you're trying to become a content strategist and ask if you can volunteer to receive a reduced rate for attendance.

STC's membership fee is well worth it, especially if you're a student. You get access to the organization's salary database, the ability to join a Slack channel with thousands of established professionals, discounted webinars, and so on (*Membership benefits*, n.d.). Plus, you can add a local STC chapter (*Communities*, n.d.), many of which will accept you regardless of where you're physically located.

Finally, you should go to things like Meetups (Meetup, n.d.) in your local area that are devoted to content strategy, if there are any. If there aren't: start one. If you're at a university, try starting a student group! The more you can network with other like-minded professionals, the more likely one of those professionals will end up hiring you someday.

Credentials Are Great, but Not Essential: Experience Is What Matters

You may have heard that top tech companies like Google and Apple are no longer requiring college degrees for their applicants (Connley, 2018). Some of our colleagues in higher education started to panic when this trend emerged, but for those of us who were following trends in fields tied to the private sector, we weren't surprised. The message these employers are trying to send is not that college doesn't matter, but rather that what matters is *what you can do* not *what you know or have studied*. For new job seekers, this kind of thinking is often anxiety-producing, however, as you may feel that you don't have enough hands-on experience. If you're trying to launch a career as a content strategist, you may wonder if your experiences in an educational, volunteer, or other work setting will matter at all.

The key to the current job market for fields like content strategy, however, is simply to demonstrate as much as possible what you can do. And yes, the best demonstration of your abilities is direct experience doing the thing you're applying for. So, you probably rightfully feel like you are in this trap represented in Figure 14.1.

We like to remind early career professionals, however, that what employers want, and what they get, are often two different things. Employers often want, well, a version of the authors of this book: someone with in-depth experience in content strategy, a client portfolio, and so on. But what they will *get* is the pool of applicants who apply to their job. That pool will most likely include many people who are relatively new to the field, especially if the job is a contract (limited term) or entry-level position.

Employers always want the moon. They want an applicant to their entry-level job who has 20 years of experience. That typically isn't the reality, however, unless the demand for jobs is extremely low or there are too many applicants to fill existing positions, which, in our experience, simply isn't the case in content strategy. On the contrary, when we talk to hiring managers and recruiters, they often tell us that it's difficult for them to find qualified candidates for their content strategy positions.

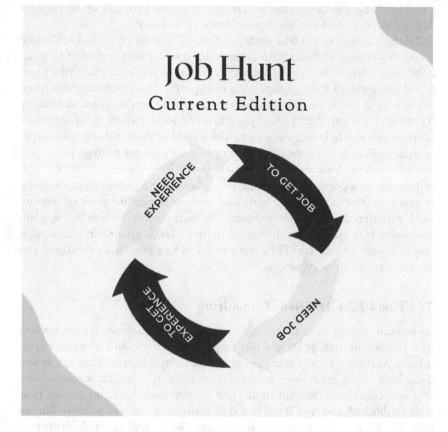

Figure 14.1 Job hunt: current edition

This means there's an opportunity for people new to content strategy to land jobs they wouldn't be able to otherwise. So our advice is: if you meet many of the minimum requirements for a job in content strategy (education-level, basic work experience, skill sets, etc.), then you should apply. Don't be intimidated by recommended requirements, because you don't know who's in the applicant pool.

But while you're applying, you can make yourself more competitive by building a portfolio.

Fake It Until You Make It: Volunteer to Build a Portfolio Until You Get On-the-Job Experience

In order to demonstrate your aptitude in content strategy: save everything you create as you learn about it. And do volunteer work in content strategy to create deliverables. We've personally seen this work many times. If you're a career-changer or a student, you're going to need to skill up in content strategy anyway. And the best way to learn is by doing.

Taking a course in content strategy? Save all of your projects. Don't have access to a course? Contact your favorite local non-profit and ask if you can help them with their content strategy. As one of the authors of this book found in a series of focus groups with non-profits, these organizations often suffer from a severe knowledge deficit when it comes to content strategy (Getto & Flanagan, 2022). Ask your favorite local charity or community center if you can help them with a content audit of their website, an analysis of their audience, and even the creation of a full content strategy plan.

Once you've done this a few times, you'll have some deliverables you can display in a portfolio. Your portfolio should be a public-facing website so that it is searchable online. It should contain things like your education, work experience, and sample projects. For each project, be sure you explain the *process* you used to develop your deliverables. Remember: that's what employers are looking for. They want to know that you understand, and can do, the content strategy process.

The Road Less Traveled: Consulting

As content strategy consultants, we also get asked sometimes if professionals new to content strategy should just forgo a regular job and become a consultant. And we typically warn people away from this path. First off, if you don't have direct experience in content strategy, why would someone hire you as a consultant? Second of all, talk to any consultant, no matter how well-established, and they'll tell you that consulting can be a feast-or-famine lifestyle. Your ability to attract and retain clients will often determine if you can eat each month, at least when you're first starting out. Plus, you're competing against all the people who have been consulting for years or even decades.

We're definitely not saying *don't be a consultant*. We, ourselves, are consultants, after all. What we are saying, however, is: start with a day job. We're still at our day jobs. The consulting agency that two of us work at, Content Garden, Inc., has never generated enough revenue to equal what we make through our day jobs. And that's ok. We're not really trying to be full-time consultants, which proves that there's value to being a part-time consultant while you work a day job.

Not all day jobs will allow this, of course. Some will make you sign non-compete agreements or other types of contracts prohibiting you from doing what you're doing for them for another organization. However, if you're in a position to do consulting, it can be very rewarding, especially if you work with smaller organizations like small businesses and non-profits, most of whom struggle with content strategy. And if and when you get established enough as a content strategist to become a full-time consultant, you will have already done all the work associated with building your portfolio, so you'll be at a different place in your career than you are right now.

The main value we find in consulting is staying current with the field. When working for a large organization, be it a university or a corporation, you are often a small cog in a very big machine. When you're a consultant, it's just you. You get to run the show. You get to advocate for whatever best practices you think are most advantageous to your client.

Another reason to do consulting is to explore new opportunities. Sometimes the best entry into a new, permanent job is to do contract work or consulting for an organization. Consulting allows you to get your feet wet with a new organization without committing to them or without them fully committing to you. If things don't go well, no harm, no foul. You finish your work with them and move on.

Choose Your Own Professional Adventure

The good and bad news is: content strategy is a wide-open field. There are lots of opportunities but this is not a field where you can rest on your laurels and expect to be successful. That's nothing new for any communication-based field, however. These fields change constantly and require you to wear a lot of hats. If that excites you, then consider becoming a content strategist (or a technical communicator or a UX designer). If it fills you with dread, this might not be the field for you.

Regardless, there are many varieties of content strategy out there. Our experiences talking to people in this field indicate that though the majority seem to use many of the skill sets we've presented in this book, their workplace environments and the specific ways they apply these skill sets differ strongly. A content strategist in the medical field is going to face very different challenges than someone working to help an education startup with their website.

There are many diverse opportunities for people interested in the skills we've discussed in this book, in other words. You may end up as a technical writer who also does content audits. Or you may be happiest as a usability expert who also does content design. Or as a technical writer who transitions to optimizing website content for SEO. These descriptions are from real people we've met. Content strategy is very interdisciplinary, meaning it draws on skills from a lot of other fields.

In fact, chances are, if you end up in a communication-related role, you will face challenges related to content. So, like usability or the ability to write concisely, content strategy is becoming a bundle of skills that are useful to many different professions. We've met lots of people who didn't start out as a content strategist, but ended up there because the organization that hired them asked them to solve content-related challenges.

Regardless, we hope this book will be useful as a reference guide and learning tool no matter where your professional destiny lies. We've tried to balance skills, stories, and exercises to appeal to a broad swath of people who are interested in this exciting field, be they researchers, teachers, students, or practitioners of content strategy. We are not certain of much when it comes to content strategy, but we are certain it will continue to be important to our increasingly global society as new forms of content are developed, published, consumed, circulated, and used for everything from growing your own food to deciding when to see a doctor.

To reference the inimitable Bill Gates, content is still king (Gates, 1996). It is as important as the technologies that support it, if not more so. We often use products and services for the *content* they make available to us, for the information we can access via them. And content is only multiplying as time goes on. It's getting more complex. And so we will need smart, capable people to continue to tame it.

We hope, if you've made it this far, you will join us on this wild adventure called content strategy.

Getting Started Guide: Launching Your Career in Content Strategy

To explore opportunities in content strategy, try doing one or more of the following:

- Change your LinkedIn profile tagline to: Seeking Opportunities in Content Strategy. Be warned: recruiters may find you!
- Identify as many people as you can on LinkedIn that are in positions in content strategy that you want to be in someday. Connect to as many of those people as the LinkedIn "People You May Know" algorithm will allow.
- When people accept your connect request, send them a message asking to do an informational interview with them. Tell them you are a student (or your current job description), you're interested in breaking into their field, and you'd like to talk to them for 15 minutes or so about how they got established as a (their job description). As a rule, you should start this process with people at the center of your current network and work outward.
- Join the groups on LinkedIn, Facebook, and Community that we reference here and introduce yourself and your career goals!
- Use LinkedIn and other search engines to look for internships and entry-level positions in content strategy. Create a spreadsheet and track all the skills sets you see repeated most often. Then skill up in those skill sets in the ways we describe in this chapter and apply to as many jobs as you feel qualified for.

- Identify 10–20 thought leaders in content strategy. Follow them on their blogs, Twitter feeds, LinkedIn feeds, and so on.
- Develop a strategy for networking in-person. Identify local in-person events related to content strategy and plan to attend. Dedicate yourself to doing this every month.
- Join STC and sign up for an event related to content strategy. Also, consider contacting the hosts of Lavacon or Confab and asking about ways to volunteer.
- Repeat for the rest of your life as needed.

References

Adler, L. (2016, February 29). *New survey reveals 85% of all jobs are filled via networking.* LinkedIn. Retrieved February 4, 2022, from www.linkedin.com/pulse/new-survey-reveals-85-all-jobs-filled-via-networking-lou-adler/

Communities. Society for Technical Communication. (n.d.). Retrieved February 17, 2022, from www.stc.org/communities/

Connley, C. (2018, October 8). *Google, Apple and 12 other companies that no longer require employees to have a college degree.* CNBC. Retrieved February 4, 2022, from www.cnbc.com/2018/08/16/15-companies-that-no-longer-require-employees-to-have-a-college-degree.html

Gates, B. (1996). *Content is king (1/3/96).* Bill Gates' web site—columns. Retrieved February 18, 2022, from http://web.archive.org/web/20010126005200/www.microsoft.com/billgates/columns/1996essay/essay960103.asp

Getto, G. & Flanagan, S. (2022). Helping content strategy: What technical Communicators can do for non-profits. *Technical Communication, 69*(1), 54–72.

Meetup. (n.d.). Retrieved February 17, 2022, from www.meetup.com/

Membership benefits. Society for Technical Communication. (n.d.). Retrieved February 17, 2022, from www.stc.org/membership/benefits

Index

Printed in the United States
by Baker & Taylor Publisher Services

Printed in the United States
by Baker & Taylor Publisher Services